X 222

JANET

Beating
the
Radar
Rap

Beating the Radar Rap

Tested techniques for fighting electronic speed entrapment—and winning

by
Dale Smith
and
John Tomerlin

Bonus Books, Inc., Chicago

94   93   92   91   90   5   4   3   2   1

Library of Congress Catalog Card Number:
89-81940

International Standard Book Number:
0-933893-89-2

Bonus Books, Inc.
160 East Illinois Street
Chicago, Illinois 60611

Printed in the United States of America

Acknowledgments

The authors gratefully acknowledge the invaluable assistance of Janice Lee, RADAR, for helping put it all together and David B. Sloan, Esq., O'Hara, Ruberg and Taylor, for his legal expertise.

CONTENTS

PREFACE

The last few years have seen some important changes in traffic laws and enforcement. The adoption of a 65 MPH speed limit on some highways, while retaining 55 on other virtually identical ones, has caused confusion and led to more than a few questionable speeding tickets. Camera radar—a device that monitors traffic and takes a picture of any vehicle that appears to be going over the speed limit—has gone into use in a few communities, and is being considered by several more. Efforts to ban radar detectors are being made by representatives of the insurance industry, which profits from surcharges on motorists who get tickets.

About the only thing that hasn't changed is the tendency of police radar to make mistakes. The few changes which have been made in radar technology have nothing to do with increasing its accuracy or reducing incorrect target identification, but are attempts to defeat radar detectors!

Win some, lose some

Since the first edition of *Beating the Radar Rap* was published, we've received many letters and phone calls from readers. Some have told us of courtroom victories, either outright dismissals or reductions in fines and points.

Others were less successful but felt they'd learned something worthwhile from the experience.

One lady telephoned to say she'd gotten a ticket moments after pulling out of the parking lot of a shopping mall. "The officer showed me a reading of 45 MPH on his radar," she reported, "but there's no way I could have been going that fast. What happened?"

We told her about phase-lock loop (PLL) radar, which can multiply a low speed reading, doubling or even tripling it. Not infrequently, an officer who hasn't been paying close attention will see such a reading, assume that it was caused by the nearest vehicle, and end up writing an unjustified ticket (see page 15).

A reader in Washington, D.C., was cited for doing 49 MPH in a 35 MPH zone, even though he knew he was innocent. He asked if the arresting officer was aware that a radar's beam spreads like the beam of a flashlight, covering several lanes of traffic after a short distance, and that it reads the speed of vehicles in both directions, either coming toward it or going away. "I think the reading you got was from that car that passed in the other direction," the driver said. "If you'll check your manual, you'll see that could have happened."

The policeman looked uncertain, but went ahead and wrote out a ticket.

Four weeks later, when the driver arrived in court to defend himself, he noticed the officer sitting a few rows away. When his case was called, he stood, but the patrolman remained seated. The judge called the case a second time, glanced around questioningly, then declared, "Since there is no one here to charge the defendant, this case is dismissed."

Apparently, the patrolman had discovered in the meantime that our reader was right!

Not all stories have a happy ending. One reader describes going to court in Nevada, and coming up against the full force and majesty of the law. "I guess I tipped them off when I began to subpoena evidence," he reports, "because the district attorney *himself* showed up to prosecute!

"I thought I could win anyway. I showed there could have been electromagnetic interference, that the reading could have been from another car, that the officer's training was below federal minimums, and that the equipment had been improperly maintained and calibrated. But the instant I finished my summation, the judge announced that I was guilty. I think it had been explained to him that I was a threat to local revenues."

Not every law enforcement agency feels threatened by *Beating the Radar Rap*. In fact, we're told that the book is used by instructors in several police-radar training schools. At least one court-operated traffic program we know of also uses it.

We're happy to help them out.

Camera radar

Many of the questions we've been asked are about devices that were mentioned only briefly in the first edition, or devices that have come on the scene since it was published. Camera radar is an example.

Camera radar is widely used in parts of Europe and Japan, but it has encountered some problems in the United States and Canada where people view a piece of equipment that takes their pictures arbitrarily and without warning as an infringement of some fairly basic rights. The device photographs not only the alleged speeder's car, its driver and anyone in the car with the driver, but also any *other* cars that happen to be present, *their* drivers and passengers.

Some people have questioned what amounts to seizure of evidence without probable cause; others have raised the issue of the lack of a charging officer or other witness to the alleged offense—an important point, considering that camera radar is subject to the same errors as other radar, plus a few of its own.

Camera radar can produce false speed readings as a result of electromagnetic interference; can result in erroneous target identification when more than one vehicle is present; and can boost a vehicle's apparent speed by as much as 10 MPH if it is changing lanes when the reading is taken.

First used in the United States in a couple of small Texas towns, camera radar soon came under fire from local citizens. One problem was the matter of "vicarious liability." This is the legal presumption that the registered owner is responsible for any violation committed in his car, whether he was operating it or not. The law in most parts of the United States (though in few other countries) prohibits such an assumption except in the case of parking tickets, and Texans seemed unwilling to see this principle changed to accommodate camera radar.

The device's promoters, Traffic Monitoring Technologies (TMT), showed up next in Paradise Valley, Arizona, where the city fathers obligingly passed a statute approving vicarious liability. This enabled them to begin collecting fines from several hundred motorists a day, the majority of whom were traveling at the 85th-percentile speed and thus posed no increased risk of accidents whatsoever.

Lately, the city of Pasadena, California, has contracted with TMT; but subject to a change in the law on vicarious liability, Pasadena authorities must content themselves with mailing out demands for money from unsuspecting motorists, and ignoring those who don't pay.

Laser speed measuring

Laser devices are being tested and may soon be in use in some localities. They were invented solely to defeat radar detectors, and suffer from several deficiencies including size, weight, cost and unreliable aiming.

There have been some fears that shooting motorists with a very narrow laser beam might be harmful to their eyes or other bodily parts; but manufacturers are confident that the devices are harmless at the energy levels involved.

Apart from such shortcomings, real or theoretical, electronics engineers maintain that they can, indeed, develop a detector that will be effective against the laser devices (almost sure to be misnamed "laser radars").

Multi-path radar

About the only new development that hasn't been motivated specifically in response to detectors is multi-path radar. As represented by _Kustom Electronics_ H.A.W.K., multi-paths are dual-antenna radars with the ability to measure the speed of targets ahead or behind the patrol car, either approaching it or receding from it.

In order to tell which of several vehicles he may be getting a speed reading from, the officer must first select the correct antenna (front or rear), choose between settings for the near or far lane of traffic, then set a switch for the direction of travel. All this, in addition to the usual procedures for moving radar, including target selection, target-speed verification, patrol-speed verification, and the use of audio doppler to obtain a tracking history.

The number of potential errors involved in the use of multi-path radar is almost impossible to calculate, but is greater than for any other type of speed-measuring device currently in use.

VASCAR and aerial clocking

The popularity of VASCAR (an acronym for Visual Average Speed Computer and Recorder) seems to be increasing. VASCAR is a time-distance calculator which depends on visual observations, and the operation of a stopwatch and computer.

Its accuracy depends on several variables including correct measurement of the range over which the target is to be timed; viewing angles and stability of the "start" and "finish" markers; and operator responses in activating the timer-calculator. Errors of as little as half a second in triggering VASCAR at the beginning or end of a measurement can convert legal speed behavior into a violation.

VASCAR can be used from a low-flying aircraft, in which case it can obtain average speeds over longer distances than on the ground. This should result in greater accuracy, though much still depends on the skill and honesty of the operator. A comprehensive analysis of VASCAR and VASCAR-Plus has been done by Kenneth Moore and is available from *JAG Engineering, P.O. Box 1727, Manassas, VA 22110*.

The war on detectors

Motorists forced to drive in the vicinity of speed traps* may benefit from a radar detector. The only function performed by a detector is to make radar patrols "visible" to motorists. Inasmuch as every major study in the past 20 years has concluded that *only a visible patrol car* has any measurable effect on traffic behavior and safety, it's no surprise that detector owners have lower accident rates than nonusers.[1]

In spite of these facts, the Insurance Institute for Highway Safety (IIHS), GEICO Insurance, and several smaller insurance companies have intensified efforts to get detectors banned. GEICO presently refuses to provide coverage to drivers who admit to using detectors, and the company has filed suit seeking to overturn a Maryland State Insurance Commission decision prohibiting this practice.

Interestingly, many of the states that are considering detector bans are the same ones that have refused to adopt a 65 MPH speed limit. Last year, the average speed of traffic, the 85th-percentile speed, and the per-mile accident rate in New York, which retains 55 as its maximum, were virtually identical to states with a 65 MPH limit. The difference was that New York, exclusive of New York City, collected more than $65 million in traffic fines last year.

If detectors are banned, the profits from enforcement could be even greater.

---

*A speed trap is any road with a posted limit substantially lower than the 85th-percentile for that road, e.g., all federal interstates, all federal-aid primaries, and most expressways and freeways, most of the time.

Roads of the rule

To a large degree, speed enforcement on United States highways is politically inspired and fiscally motivated. The now dual federal speed limits are based on no engineering concepts and embody no safety principles; they were promulgated in response to an oil shortage and threatened loss of tax revenues, respectively; today, they are almost universally ignored, serving no useful purpose apart from propaganda and fund raising.

The safest roads in the world are no longer those of the United States. In 1988–89, the mileage death rate on United States interstates was 1.1 per 100 million vehicle miles; on the West German autobahns, despite the general absence of speed limits and an 85th-percentile of over 82 MPH, the fatality rate was 0.9.

Actual travel speeds are continuing to rise in the United States, as they have done for the past half century, reflecting the steady improvements in vehicles, roads and safety systems. In 1988, the 85th-percentile speed on United States interstates was approximately 70 MPH, while the mileage death rate on all types of roads dropped to 2.4, the lowest level in history. Both of these trends are likely to continue.

For American motorists caught between the mutually exclusive choice of driving at the safest speed, *i.e.*, the 85th-percentile, or driving at the *legal* speed (at least 10 MPH slower), here are a few pointers that may help:

1. Stay out of the passing lane except to pass.

Police officers, who are required to write a certain number of tickets per shift (yes, Virginia, there is a quota system), look for cars in the No. 1 lane on the grounds that, since everybody is breaking the speed limit, they might as well ticket those who are breaking it the most. Cars in the No. 2 or 3 lanes can go just as fast without drawing attention to themselves.

2. Take advantage of "rabbits."

When driving cross-country, wait for a faster motorist and tail him. Stay at least a quarter of a mile back, and half a mile if visibility permits. Chances are your rabbit knows the territory, and where to look for stake outs, but if not, he'll get stopped, not you. If you lose your rabbit, drop back to the speed limit and be patient, another one will be along in a few miles.

3. Fight technology with technology.

Radar detectors, scanners and CB radios can save you money. As long as radar is used without performance standards, without minimum training requirements, and without restrictions on speed trapping, you at least have the right to know when you're under surveillance. Warning: the use of radar detectors is illegal in Connecticut, Virginia and Washington, D.C. Radar jammers are illegal everywhere by federal statute.

4. Fight unjustified speeding tickets!

The State of California considers radar a speed trap when used to enforce limits lower than the 85th-percentile (that's one reason the CHP doesn't use radar on freeways or interstates). We agree with the thinking behind this policy. If you get a ticket while traveling with the flow of traffic—or on open highway where the posted limit is clearly unreasonable—you have been made the victim of a legally sanctioned speed trap, and you have the right to defend yourself. All you need is the will.

This book will show you the way.

John Tomerlin
Laguna Beach

SPEED, SAFETY AND RADAR: SOME SURPRISING RELATIONSHIPS

T his book is for safe drivers. It will be of little use to compulsive speeders, hostile and aggressive drivers, or those who drink without regard to proper limits, personal and legal. Drivers such as these can expect to go on collecting traffic tickets and accidents until they lose their licenses or their lives.

It will also be of little use to motorists who persist in going much slower than surrounding traffic. Federal studies show that cars traveling 5 MPH below the mean speed of traffic are two and one half times more likely to be involved in a fatal crash as are those going 5 MPH above it; at 10 MPH below the mean (55 MPH in an area where most traffic is moving 65 MPH, for example), the likelihood of having such an accident is *six times greater*.[2] Of course, slow drivers seldom are required to defend themselves in court for driving at dangerous speeds.

Speed and safety

Ironically, most speeding citations are given for speeds that are neither extremely high nor dangerously low. Most are issued to motorists driving at or slightly above the mean speed of traffic—the very speeds at which the fewest accidents occur!

This peculiar state of affairs is due largely to Congress's decision, in 1974, to set 55 MPH as the maximum speed limit for all parts of the country. Until then, a growing number of professional traffic engineers had relied on the 85th percentile (the speed at or below which 85 out of every 100 cars travel in the absence of posted limits) to determine optimum speeds.

Studies indicate that when limits reflect the 85th percentile, or "representative" speed of traffic, several good things happen. First, voluntary compliance is increased, allowing enforcement agencies to concentrate on the most serious offenders—drunk drivers and 95th percentile and above speeders.

Second, speed variance (the difference between the fastest and slowest traffic) is decreased; and speed variance *is* directly correlated to accidents.

Third, when speed limits are changed to reflect the 85th percentile, the accident rate decreases, whether the new speed is *higher or lower* than the old one.

Unfortunately, the benefits of the 85th percentile have been sacrificed to the 55 MPH national maximum speed limit. Nor does the increase to 65 MPH on some rural interstates help matters much. This move was so long in coming (where it *has* come—several states in the northeastern part of the country are clinging to 55 MPH at this writing) that the 85th percentile speed on these roads is already close to 70 MPH.

What 55/65 hasn't changed, and cannot change, is the basic process through which speed behavior is determined. The underlying principle of this process, a sort of Golden Rule for achieving enforceable speed limits, states: **"Most motorists tend to drive at speeds they consider safe and reasonable, and to ignore posted limits which they consider either too high or too low."**[3] This common sense behavior has resulted in an evolutionary relationship between technological development and speed, a relationship that accounts for the steady decline in the mileage death rate during a half century of rising highway speeds.

Speed limits and the public

Some authorities continue to argue, against virtually all scientific evidence, that 55 has resulted in "significant" savings in fuel and lives. They ignore, or try to minimize, a far more apparent result. Whereas about 15 percent of the public exceeded the speed limit before the law as passed, more than 90 percent does so on some highways today.[4] This imbalance between mandated "speed morality" and the public's view of safe and reasonable behavior has created some winners and some losers. The winners include:

- *Safety professionals.* Federal and state administrators; publicly supported safety agencies; and self-appointed (usually insurance-industry supported) "safety advocates" have all benefited from the attempt to impose a reasonable sounding, though impracticable, speed limit.

- *The auto insurance industry.* Entitled by law to raise premiums to motorists who get speeding tickets, insurers have enjoyed an unprecedented bonanza since Congress redefined what was legal without regard for what was safe. The industry underwrites its own lobbying and public relations office, the Insurance Institute for Highway Safety (IIHS), to spread misinformation about 55 and other putative safety measures.

- *State and local governments.* Enormous increases in revenues since the passage of the national maximum limits, in conjunction with federally supplied radar guns, have made traffic enforcement an important funding source for state and local governments. Many jurisdictions now budget a year in advance for money from traffic fines, relying on such revenues for a major part of their operating expenses.

Those are some of the winners. The losers include the motoring public and highway safety in general.

Radar and enforcement

It would be difficult to overestimate the importance of radar to enforcement since 1974. Without it, the national speed limit would have become a dead letter soon after normal fuel supplies were restored following the Arab oil embargo. With radar, and with tens of millions of dollars in federal grants, enforcing 55 not only wasn't burdensome to the states, it became a multi-billion dollar source of new revenues.

Almost overlooked in the rush to equip the states with radar was the fact that speed enforcement in general—and radar in particular—plays virtually no role in traffic safety. According to the Federal Highway Administration (FHwA), based on one of the largest accident surveys ever performed, **"the level of enforcement has little or no apparent effect on the mean speed (of traffic) or on . . . accident experience."**[5]

Another example of how ineffective enforcement alone can be took place in 1982, when the California Highway Patrol ran some 40,000 additional missions on four interstate highways in an effort to reduce noncompliance with 55. The result was a reduction of less than 4 percent (from 78.2 percent, to 74.7 percent noncompliance), and according to a CHP spokesman, the effects of the campaign diminished "within one or two days" after the patrols were discontinued.[6]

The only practical function of police radar is to let officers write more tickets. Less time is lost in pacing suspected offenders, and fewer motorists are forewarned of the presence of a patrol vehicle in time to change their speed. If radar is used where there is a large differential between the legal limit and the 85th-percentile speed, an officer can issue citations literally as fast as he can write them.

Most people employed in enforcement, or in the safety establishment, assume that there is a benefit *per se* in issuing tickets. Either the *threat* of a ticket deters people from some innate, unexplained urge to drive faster than is safe for themselves or others; or getting a ticket reforms the offender. Unfortunately for this theory, it is refuted by every major study of traffic behavior.

- When the National Academy of Sciences (NAS) studied data from all 50 states for the years 1974 to 1983, it found no positive relationship between the level of enforcement and either speeds or accident rates.[7]

- After the FHwA performed the largest accident study in history, it concluded that, "the level of enforcement has little or no apparent effect on the mean speed or on . . . accident experience in the study sample."[8]

- Still another FHwA study found that, while a visible patrol car is capable of influencing traffic behavior to some degree, there was no consensus on the range or duration of this so-called "halo" effect. (Some of the federal government's Selective Traffic Enforcement Programs in the late 1970s and early 1980s concluded that the halo effect reached as far as three to five miles; but other researchers have placed the range at less than one-half mile.)[9]

- The General Accounting Office (GAO) reported to Congress in 1988 that "states that aggressively ticket speeders do not necessarily motivate motorists to comply with the 55 MPH speed limit." Noting that in 1985 the state of Maryland had issued more than 180 tickets per mile of highway—between two and thirteen times the rate of other states audited by the GAO—Maryland still was unable to stay in compliance. "In general, we found little relationship between . . . enforcement efforts and (actual) highway travel speeds," the watchdog agency concluded.[10]

Radar, especially when used from places of concealment, lacks even the modest benefits of the halo effect. The only way that radar enforcement exerts even a temporary effect on traffic behavior is when its use is accompanied by large amounts of media publicity. In such cases, a modest reduction in traffic speeds (but little or no change in accidents) may occur.

A study by the University of North Carolina's prestigious Highway Safety Research Center entitled *Radar As a Speed Deterrent* found that radar alone "reflects only marginal evidence of effectiveness."[11]

Other problems with radar

If the only objection to radar was its failure to promote safety, it could be viewed as a kind of selective "highway users tax." Unfortunately, this isn't the only problem. In the rush to get radar into the hands of as many enforcement agencies in as short a time as possible, the National Highway Traffic Safety Administration (NHTSA), and almost all state legislatures, brushed aside the need for acceptable performance standards. Proposed federal standards for equipment performance and officer training were dropped when it was discovered that they would have made virtually every radar in use at the time obsolete, as well as disqualifying their operators.

To date, only a handful of states has enacted any form of radar regulations, and none of these are rigorous enough to disqualify even the poorest and most inaccurate radar equipment presently in use. As a result, a substantial number of the estimated 15 million radar-backed speeding tickets issued nationwide each year are for either incorrect speed readings, or mistaken vehicle identification. The actual proportion of unjustified tickets is impossible to know, though reliable observers have suggested it could be in the range of 10 to 12 percent, overall, and substantially higher for moving radar.

Such figures are far higher than they should be. As Judge Alfred Nesbitt noted in his decision *Florida v. Aquilera*: "For the average law-abiding American citizen, minor traffic offenses constitute the only contact such a person will have with the law enforcement and judicial systems. Public confidence relies upon the fairness of such proceedings."

Today, the "average law-abiding American citizen" is under siege. Forced to choose between the safest speed or the legal speed; subject to enforcement policies that are either irrelevant or counterproductive; victimized by insurance companies seeking ever higher premiums—the average citizen has only two lines of recourse. The ballot box and the courtroom.

At the ballot box, you have the right and the opportunity to vote for informed, responsible highway safety policies, and against candidates who permit themselves to espouse uninformed opinion on what is, after all, one of the nation's most serious health problems.

In the courtroom, you have the privilege of defending yourself against unwarranted traffic tickets and improper enforcement policies. You have the right to question the reliability and accuracy of the radar unit; the circumstances under which it was used; even the rationale behind arbitrary and unrealistic speed limits.

In availing yourself of every legal safeguard to which you are entitled, you can force the system to confront, and perhaps recognize, some of the more careless and complacent assumptions it makes about highway safety.

Chapter One

WHY FIGHT IT?

So long as the authorities use enforcement for fund raising, you have a right to fight.

N o one really likes traffic court. Judges fight to avoid being assigned to it. Hollywood doesn't make movies about it. Until a few years ago public opinion was about evenly divided on which was more unpleasant, traffic court or oral surgery. Since the invention of high-speed drills and improved anesthetics, traffic court appears to have the edge.

The authorities aren't concerned by this. You don't hear complaints about "declining public interest in our traffic court system." Instead, there has been a systematic effort to get accused speeders to plead guilty and forward a token of their penance by mail—no need to become personally involved at all.

Those who insist on having their day in court may find themselves shunted off to a "traffic adjudicator" (basically a glorified cashier); or told that their transgression is so unimportant—only an "infraction," really—that they aren't entitled to a trial by jury; or faced with bail higher than the cost of settling out of court. All of which can be discouraging, and plainly is intended to be.

Some motorists decline to contest traffic tickets on the basis that, since they've exceeded the limits on other occasions and not been caught, they're merely receiving delayed justice. It doesn't seem to occur to them that the limits on most highways today are set for political, not safety, reasons and that they may not have *deserved* a ticket on those earlier occasions either.

Another common reason for not fighting is the time and effort involved. How much does a ticket cost, anyway, $25 to $40? You lose more than that from a day off work, not to mention the trouble of preparing a defense. Cheaper to pay the "radar tax" and forget it, right?

Well, not always. Not if you've had a ticket within the past three years, for instance, in which case most insurance companies will use the next one as an excuse to hike your premiums several hundred dollars. Not if you're a high-mileage driver with fair prospects of getting a ticket from time to time; and not if you're a professional driver whose license is in jeopardy as a result of speed traps.

In these and similar cases, fighting—and winning— may be vitally important.

Two other reasons, seldom mentioned, for not challenging a ticket are the fear of courts and the fear of losing. For most of us, courtrooms are unpleasant places, stiff with the formality of the law and bleak with human sadness. To escape from such a place seems enough to hope for, never mind being found innocent. Even if we are guiltless, the officer's word backed by radar evidence makes conviction a certainty, and if we really were exceeding the limit, no legitimate defense is possible—so why bother?

The answer is that both of these notions are false. Police radar can and *does* make errors; and thousands of miles of U.S. highways are posted at arbitrary and unrealistic speeds, making them in effect speed traps. In both circumstances you are legally and morally entitled to a defense.

How reliable is radar?

Radar was developed during World War II to locate and track enemy targets (the name is a quasi-acronym from Radio Detection and Ranging), a job at which it was highly effective. An early example of "glamor technology," radar was adopted into police work like a returning war hero, and for many years was considered all but infallible by most members of the public. Radar evidence was somewhat slower winning acceptance by the courts, but by the mid-1960s had gained "judicial notice" (a legal ruling establishing certain evidence as beyond dispute).

The misconception embedded in this view is that military/aviation radar and the police traffic variety are essentially the same. In truth, about the only similarity between the two is that both use microwave radiation to seek their targets.

In terms of sophistication, the difference is vast. Typically, police radar units sell for between $2,000 and $3,500 (though one of the most popular makes in use sold in large quantities for under $400), while commercial airport radars and the Defense Early Warning system (DEW line) are multi-million dollar installations. Comparing the two kinds of radar units is like equating a Roman candle with an ICBM.

The Doppler shift

All types of radar rely on a principle first described by Austrian physicist, Johann Doppler, which states that waves striking a moving object will be reflected to their source at a different frequency, dependent on the speed and direction the object is moving. Doppler had sound waves in mind—as in the famous illustration of the changing pitch of a train's whistle—but the Doppler shift is apparent in any form of radiated energy.

As shown in Figure 1, a microwave signal leaving a radar antenna resembles an expanding cone, like the beam from a flashlight. The beam grows wider, and weaker, as it propagates; at 1,500 feet, it is wide enough to cover all four lanes of a divided highway, while at one mile (5,280 feet), part of the beam will cover an area wide enough to land and take off a small plane.

In Figure 2 the beam has encountered a target, an automobile moving toward the radar at an undetermined speed (the beam itself moves at the speed of light, 186,000 miles/second, and at a fixed frequency, or wavelength). When part of the beam is reflected from the target, this "echo" returns to the radar at a slightly different frequency. The difference between the frequency transmitted and the one received back can be measured and converted into an equivalent number of miles per hour.*

Note that the radar's computer only measures the *difference* in the two signals, not whether one is higher or lower than the other. In practical terms, this means the radar cannot tell whether the target is approaching or retreating, only the speed at which it is doing one or the other.

---

*31.4 cycles/second for X-band, and 72.05 cycles/second for K-band, equals one MPH, in case you're counting.

FIGURE 1
The Width of a Radar Beam

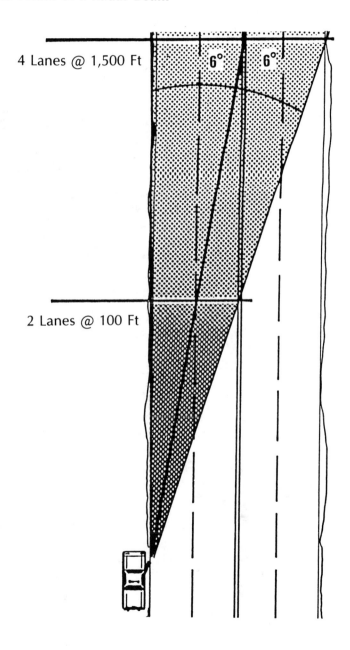

4 Lanes @ 1,500 Ft

6° 6°

2 Lanes @ 100 Ft

FIGURE 2
The Doppler Effect

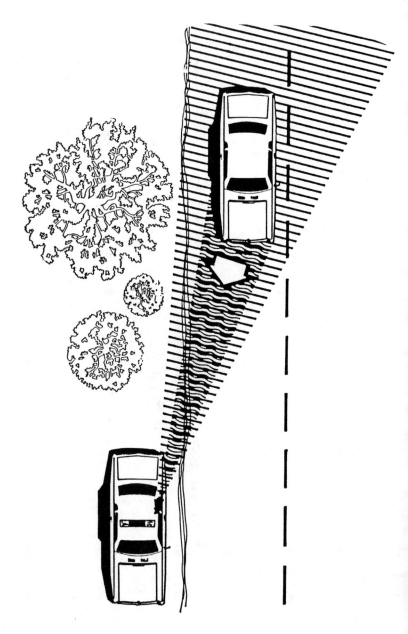

Phase-lock loop (PLL) radar

To make matters worse, modern traffic radar no longer relies on direct measurement of frequency changes to determine target speed; instead, a circuit known as a phase-lock loop (PLL) is used to *interpret* the microwave shift and display it as a digital readout in miles per hour.

The advantages of PLL units are smaller size, lower cost, and the ability to read a target more easily in the presence of "noise" than true Doppler radars.

Disadvantages include excessive range (up to a mile and a half for some models) and a pronounced tendency to lock on the first good reflective surface in the neighborhood. Whereas true Doppler radar measures the frequency shift in a returning signal several times—recycling itself if one of the readings varies for any reason, and continuing until the reading is constant—PLLs must sort from a high noise-to-signal mixture to begin with, giving rise to a wide variety of possible "ghost" readings.

PLL radar has a marked predilection for "harmonic error," doubling or tripling the speed reading of vehicles that are moving 20 MPH or less. If a patrol car equipped with PLL radar is changing speed, and if the target vehicle is accelerating from a low speed (as when exiting a driveway or parking lot, for example), a true speed of, say, 18 MPH might show up on the radar as 36 MPH or 54 MPH—either of which may be above the posted limit. If the officer isn't observing traffic carefully, or has obtained a reading at some distance, he may issue an unjustified ticket.

Notwithstanding all of the above, current versions of police radar can measure the speed of a moving vehicle with great precision, provided that (1) the road is straight, (2) visibility is good, (3) traffic is light, and (4) the officer has been properly trained to identify and ignore all spurious signals. In practice, of course, conditions are seldom this ideal and errors both in speed readings and target identification are everyday occurrences.

Radar and the law

Ironically, the first highway application for radar was of great potential benefit to motorists. In 1946, traffic engineers used it to study traffic patterns and set speed limits to reflect consensus. If it had continued to be used for this purpose alone, radar would be a valuable tool of traffic safety.

But enforcement agencies lost little time in adapting the new technology to their purposes. By the following year, the police in Glastonbury, Connecticut, were snaring speeders, using S-band radar with a readout on a strip chart. The chart, which resembled a polygraph or EKG printout, could be introduced as evidence in court, with the officer testifying that such-and-such a squiggle represented the defendant's vehicle. This physical "proof" was powerful support for the officer's charge of speeding, a charge judges had been accustomed to accepting largely on faith.

Not all jurists were enchanted by the magic box. In *People of New York v. Offermann*, in 1953, the judge denied radar judicial notice and limited it to use as corroborating evidence. But the following year, also in New York, *Buffalo v. Beck* established criteria for judicial notice on the basis that knowledge of electronics wasn't necessary in order for an officer to use radar, or for the court to accept it as evidence.

In 1955, *People v. Sachs* expanded the requirements for judicial notice by demanding proof that "the radar was properly set up, that its instruments were working, that the apprehending automobile was set in its own location (in other words, the police car had to be stationary), and that both cars were visible to each other at a reasonable distance." *Sachs* also required radar to be "checked with a calibrated speedometer at the beginning and end of the tour," that "the strip charts show the results of tests," and that the "radar officers observed the speeding vehicle."

These cases were interesting chiefly as opening shots in the legal battles that would ensue.*

The first of three landmark cases, on which much of the legal status of radar currently rests, was *New Jersey v. Dantonio* in 1955. Here, judicial notice was taken of the Doppler Principle, and it was established that *no technical expertise was required to operate the equipment.* Although the decision in *Dantonio* specified that the equipment should be properly set up, no definition of a proper setup was given; and the calibration drive through the radar's beam was deemed hearsay unless the officer who drove the second car also testified. A strip chart continued to be necessary as physical evidence of the radar's findings.

Little else of significance happened until 1965, when the cumbersome S-band radar was replaced by X-band. In addition to being smaller, hence more portable, X-band dispensed with the strip chart, substituting an analog speed meter with a needle moving across a miles-per-hour scale. In 1966, *Connecticut v. Tomanelli* honored X-band by changing the test requirements for radar. It established that the use of a tuning fork was "sufficient to verify that the radar was operating properly and producing correct speed readings."

The third case that contributed to judicial notice of radar was *Kentucky v. Honeycutt* in 1966. In addition to reaffirming that radar operators require no special train-

---

*Cases of interest are outlined in Appendix A.

ing, *Honeycutt* established that the vehicle responsible for a reading is "the one that is out front and closest to the radar."

What these decisions had in common is that, factually and scientifically, all three were wrong. Contrary to *Dantonio*, radar operators do indeed require special training if mistakes are to be avoided; in spite of *Tomanelli*, a test with a tuning fork is *not* sufficient to prove that radar is operating properly and producing correct speed readings; and *Honeycutt* notwithstanding, the vehicle out front and closest to the radar is by no means necessarily responsible for a particular speed reading. But it wasn't until thirteen years after the *Honeycutt* decision that the infallibility of radar was severely tested in court.

The Miami radar trial

In 1974, after the 55 MPH speed limit was passed, the National Highway Traffic Safety Administration began purchasing radar for the states. By 1978, approximately 12,000 federally funded units were in use, producing between six and seven million tickets per year. Not only were the majority of these units inferior in performance, virtually all were being used by untrained or insufficiently trained officers.[12] Sooner or later, complaints from unjustly accused motorists were bound to surface.

After five years of what can only be considered a conspiracy of silence by federal, state and local authorities, the truth about radar's weaknesses began to surface. This was largely due to the efforts of a crusading television station and a courageous municipal court judge in Miami, Florida.

Chapter Two

RADAR AND THE LAW

Why electronic speed measurement is not infallible.

I n February of 1979, viewers of WTVJ-TV watched in fascination as a popular make of police radar clocked a palm tree supposedly moving 86 MPH, and a house ambling along at a more stately 28 MPH. Both readings were spurious, of course, produced by external interference rather than the actual targets,* but they served to demonstrate how radar can, and sometimes does, go wrong.

One of the persons viewing the program was Judge Alfred Nesbitt of the Dade County Circuit Court. Nesbitt, who was scheduled to preside over several cases involving radar, notified the local prosecutor and the public defender's office that he would consolidate the cases and receive detailed arguments for and against the accuracy of radar speed measuring devices.

The result was *Florida v. Aquilera*, a case that would become known as the "Miami Radar Trial" and make headlines across the nation.

During the proceedings, expert witnesses from the electronics field and radar-related industries testified to a variety of potential errors that could be made by radar. The

---

*Panning the antenna, in one case, and the fan motor in the patrol car in another.

public learned for the first time that there were *no equipment standards* for police radar, and *no training requirements* for the officers who used it.

Wide discrepancies continue to exist today in the quality and performance of different makes of radar, and in the training programs for their use.

After nine days of detailed testimony, Judge Nesbitt reached a conclusion that had far-reaching implications. In dismissing the charges against Anna Aquilera and 80 other radar defendants, Nesbitt noted that the "reliability of the speed measuring devices as used in their present mode...*has not been established beyond...every reasonable doubt*" (emphasis added).

He went on to stress the need for adequate training both in the classroom and in the field, and examinations to determine the operator's qualifications. "Let it be understood once and for all," Nesbitt concluded, "that the function of the traffic court is to convict the guilty, acquit the innocent, and improve traffic safety, not to be merely an arm of any revenue collecting office." (The case is described in greater detail in the next chapter.)

What mistakes does radar make?

There's little agreement on how often radar is wrong or how many arrests are made on the basis of incorrect readings. Evidence presented in Miami suggested that in the case of *moving* radar, where target identification time is shortened, and background radiation may escape detection, the figure could be as high as 25 percent. A more conservative estimate is supplied by Col. Lee Nichols, head of the department of electrical engineering at the Virginia Military Institute, who believes that erroneous citations occur no more than 10 percent of the time.[13]

"Either figure is higher than it should be, and higher than it needs to be," Nichols has said.

Under ideal circumstances—with high-grade, properly maintained and checked equipment in the hands of an officer able to identify and avoid every likely source of interference—the probability of a ticket being issued in error is near zero. To achieve these conditions, however, traffic must be moderate to light, to insure target separation; road conditions must be good, to encourage normal speed behavior; and visibility must be clear for over a mile (in daylight and fair weather), to avoid target misidentification.

This is a good description of conditions where little or no enforcement is needed.

Assuming that the purpose of radar is to promote safety, it would seem that the closer it comes to achieving zero errors, the closer it comes to providing zero benefits.

Speed traps

In actual practice, radar makes a significant number of errors, while producing few if any safety benefits. This is especially true of speed traps, which may be defined as any enforcement activity whose primary result is the writing of tickets rather than the reduction of accidents. This simple definition is confounded by the notion that, if money can be raised at the same time safety is enhanced, so much the better—an attitude which too often leads to financial objectives taking precedence over safety.

Many small communities have made a practice of posting abnormally low limits for the sole purpose of raising revenues. For a time, Selma, Texas, collected 80 percent of its annual budget from speeding tickets; and Friendly, West Virginia, with a population of 190, collected $3,100 in a single two-month period—until their respective state legislatures forced them to stop, or at least cut back.[14]

In 1983, some communities in Indiana lowered their posted limits to 29 MPH in residential areas, 20 MPH in business districts, and 10 MPH in school zones for the express purpose of increasing revenues. Leaders in one town said they hoped to raise an additional $40,000 per year, without even the pretext of improving safety.[15]

Most such abuses stem from mixing money from traffic fines with other public funds. Once it's been laundered, the cash may be spent on a variety of non-safety-related items, not excluding salaries and benefits for those who create and administer the laws.

Common sense as much as common decency suggests that while accidents still occur, and citizens are being killed and injured by them, all money collected from traffic offenses should be spent to increase road safety— provide more and improved enforcement, lighting, grade-crossings, crash attenuators, median barriers, etc. To spend it otherwise either shortchanges safety—in which case citizens are being killed and injured unnecessarily—or denies that there is a safety problem, in which case traffic fines amount to armed extortion.

State laws should, and in some cases do, prohibit the mingling of traffic fines with general funds; but such restrictions rarely are applied, one reason being that the states themselves are deeply involved in profiting from enforcement.

In California, for example, where legislation specifically forbids speed traps—and defines a speed trap as any form of point-to-point timing—aerial speed measurement is routinely used by the highway patrol.*

A California law requiring radar to be used *only* where speed limits are determined by the 85th-percentile speed is ignored in order to carry out technically illegal aircraft surveillance.

In New York, which kept 55 MPH as the limit after most states had gone to 65, the state collected over $60 million in traffic fines during 1988 exclusive of New York City.

Encouraged by the NHTSA, the National Safety Council and other members of the safety establishment—most of whom are subsidized directly or indirectly by the insurance industry—state and local officials apparently see nothing wrong in making traffic enforcement the servant of fiscal objectives.

What if I was speeding?

None of these observations is meant to excuse the act of driving at dangerous speeds. An automobile can be a deadly weapon, and those who use it in an overtly hostile and aggressive way should be subject to the full sanctions of the law. The question is, when is speed alone dangerous?

Over the years, there have been many attempts to establish the precise relationship between speed and safety. The common perception that speed is hazardous *per se* appears to be based on an innate fear of acceleration. Which may explain why, in an age of supersonic travel,

---

*Contrary to popular belief, there is no such thing as aerial police radar; all measurements are for distance/time between fixed ground references.

the National Safety Council is able to maintain that "speed kills" without exciting ridicule.

Perhaps this explains why police officers so often list speed as the primary causal factor on accident reports. Lacking any formal training in such matters, the average officer feels that the *consequences* would have been less if the crash had occurred at a lower speed. Which is true, of course, but irrelevant to the question of causation.

The total energy involved in a collision increases exponentially with speed, so that an accident at 70 MPH is 1.62 times as destructive as one at 55 MPH. Of course, 55 isn't safe to crash at either—about 80 percent of all fatalities occur at 40 MPH or under—meaning that the key to greater safety lies in reducing the causes of accidents, not just the speeds at which they happen.

Is speed a major cause of accidents? Every major traffic study since World War II has shown that it is not. In the late 1940s, before all of the states adopted formal speed limits, Matthew Sielski, a traffic engineer for the Michigan Automobile Club, analyzed crashes on several Midwestern highways and found no significant difference in the rate of fatal accidents on either side of state lines, where one of the states had a posted limit and the other hadn't. He also found that speed "too fast for conditions" (not merely in excess of posted limits) could be considered the primary cause of only 4 percent of all fatal accidents.[16]

The largest such study ever made was an analysis of more than 6,000 fatal crashes, by David Solomon of the Federal Highway Administration (FHwA). First performed with data from main rural highways, and later repeated with data from the Interstate system, both the initial study and the followup showed the speed at which the fewest fatal accidents occur is about 5–10 MPH above the average speed for all traffic.[17]

They also showed that the actual speeds motorists travel are largely independent of either posted limits or the level of enforcement.

It may be understood from all of this that any highway posted at 55 MPH—where 55 is *not* the 85th-percentile speed—is a speed trap by definition. To oppose such an abuse of the public's confidence by fighting unjustified speeding tickets is more than a right, or even a privilege, it's a duty.

What are my chances of winning?

It's never been an easy matter to beat a traffic ticket. You may obtain a slightly lower fine by pleading (the term is apt) "guilty with an explanation," but this is more a reward for sincerity or imagination than an exception to the rule.

The rule seems to be: in traffic court you're guilty until proven innocent, and sometimes even then.

Usually it's a matter of your word against the officer's; and if radar was used, most judges will accept this as independent—certainly impartial, possibly infallible— evidence of guilt. If you do nothing but stand up and declare your innocence, or express doubts about the reliability of police radar, your chances of winning are nil. Judges customarily defer to the officer's testimony, and there is judicial precedent for treating *uncontested* radar evidence as proof of guilt.

The key word here is "uncontested." For, while it may be difficult to challenge the officer's word directly, or refute the electronic principles on which radar works, it is entirely possible to create reasonable doubt about the circumstances or conditions under which a speed reading was obtained.

Granted, there may be instances where this strategy is inadequate. If you are far from home, and can't afford to drive back for arraignment—or once for arraignment and again for trial; or if you're caught up in an obvious ticket mill, where chances of obtaining justice are lower than a snake's goatee—you may have to pay the radar tax, and choose another day to fight.

If, on the other hand, your case comes before any of the overwhelming majority of honest men and women who serve on the nation's benches, you can expect a full and fair hearing, and at least a reasonable chance of acquittal.

The information in this book is offered as a general guide and is not intended to cover all the vagaries of the law in all 50 states. If you intend to conduct your own defense, you should consult the vehicle code and rules of courtroom procedure for your particular jurisdiction. Both are available through your local library.

Your success will depend on how persuaded a particular judge is of the reliability of police radar, and how effective you or your attorney are at challenging incorrect assumptions. But if you're willing to take the time to prepare yourself—to learn how radar works, and why it doesn't always work properly; how operators make mistakes, and how to get them to admit it; how to present your case, while taking full advantage of the opportunities the law provides for your defense—chances are you can join the growing number of motorists who have learned to beat the radar rap.

Chapter Three

RADAR ON TRIAL

What the Miami Radar Trial and the tests
by the National Bureau of Standards mean
to you.

T he news that radar makes mistakes should have
stirred a revolution. Armies of outraged motorists
should have marched on their state legislatures, demanding repeal of laws that made radar *prima facie* proof
of speeding. Class action suits should have been filed
against insurance companies for assessing illegal premiums on the thousands who had been ticketed but were innocent of speeding. Heads should have rolled at NHTSA
for misappropriation of public funds: the millions spent to
put inferior equipment into the hands of inadequately
trained officers.

None of this happened.

Perhaps it was because the whole business seemed a
bit hard to believe. Surely police radar couldn't be tricked
into producing virtually any reading desired, could it? Federal and state authorities wouldn't permit radar to be used
in the total absence of performance standards, would
they? Enforcement officials couldn't allow untrained or
minimally trained officers to use radar against the public,
could they?

As a matter of fact, they could.

Radar on the move

For about eight years—from the time of *Tomanelli* and *Honeycutt*, in 1966, to the passage of the 55 MPH speed limit, in 1974—there was a period of relative tranquility. With speed limits set at the 85th percentile, the great majority of motorists drove within the law, and the few who didn't could be spotted and apprehended by conventional means. Radar continued to gain in popularity, but rather slowly.

Behind the scenes, changes were taking place: design variations and performance differences that were neither recognized by the courts nor reflected in the laws were showing up on police radar.

One of those changes, the switch from true Doppler to phase-lock loop (PLL) receivers, already has been mentioned. In PLL devices, both the speed readout and the audio tone of the radar are driven by an internal oscillator which *interprets* the Doppler signal without actually measuring it. While this system works well enough against an isolated target, it is apt to become confused when several targets are within its range.

This problem was compounded as manufacturers began building greater range into their equipment. Possibly in hopes of defeating radar detectors, some major suppliers claimed a range of up to *a mile and a half*—or about four times the maximum distance at which visual identification can be made. This resulted in even more chances for error by the operators of PLL radars.

Another potentially serious source of error was introduced with "instant-on" radar. This is a feature that places the unit on standby, emitting no signal until triggered by the operator. The objective was to defeat detectors by concealing the patrolman's presence until the last possible moment; but the actual effect was to deprive the officer of a "tracking history"—an ongoing record of vehicle speeds and possible sources of electromagnetic interference.

Without such a history, there is no way for the operator to be certain that the speed reading he obtains belongs to a particular vehicle, or that the reading is free from interference.

Instant-on radar can be quite effective on lightly traveled sections of highway that have been checked for interference, which simply means it's very good for speed traps.

The greatest problems with correct speed readings and reliable target identification accompanied the introduction, in 1972, of moving radar. Whereas stationary radars actually measure the speed of a target, moving radars *compute* it—a subtle difference but one fraught with possibilities for error.

Stationary radars simply look for the strongest reflected signal and calculate a speed from the Doppler shift of that signal. Moving radar, however, must measure two parts of the same signal. One, the so-called "low beam," is reflected from the landscape (buildings, billboards, guard rails, parked cars, etc.), and represents the patrol car's own speed. The other, the so-called "high beam," is assumed to be a reflection from the target. A computer in the radar subtracts the low-beam component from the high-beam component and displays the difference as the target speed in miles per hour. This means that if the radar gets a stronger low-beam reflection from some object other than the road, it will misread the patrol car speed and subtract the wrong amount from the target speed. This almost invariably results in a higher reading for the target.

Despite its shortcomings, moving radar was eagerly received by enforcement officials, who saw in it a way to cover more territory with available resources, and, not incidentally, to screen larger numbers of vehicles for possible speeding violations.

Some members of the public were less enthusiastic. A few even decided to challenge the new technology in court. An early case was *Ohio v. Wilcox*, in which the Court, while reaffirming judicial notice for the Doppler Principle itself, noted that it did "not apply to a device which not only measures speed, but adjusts such speed measurement to compensate for the speed of the vehicle in which it is mounted."

This decision—scientifically impeccable, and prophetic of future developments—was overturned just two years later in *Ohio v. Shelt*, where it was determined that judicial notice *did* apply, not only to the Kustom MR-7 used against the defendant, but to all moving radar!

In 1978, the Wisconsin Supreme Court restored a degree of sanity to the discussion by concluding, in *Wisconsin v. Hanson*, that judicial notice may *not* be taken of the MR-7, and that moving radar could be used only "in areas where road conditions are such that there is a minimum possibility of distortion" (whatever such conditions might be).*

From this decision also came requirements that the patrol car's speed must be verified by its own speedometer; that the speedometer must be tested on a regular basis; and that testing of the radar must be done by means independent of the device's own internal calibration system.

So far, so good. At last a court had made an effort to limit some of the causes of undeserved radar tickets. It hadn't identified every source of mechanical error, or even the most important ones, nor had it even begun to address the issue of operator error; but it had confirmed that radar

---

*The benefits of this decision were lessened when the manufacturer of the MR-7 changed the color of the case, re-named the unit "MR-9," and resumed selling it to Wisconsin enforcement agencies.

could make mistakes and, by implication, that some motorists were being cited unjustly.

It remained for *Florida v. Aquilera* to show just how
many.

Spurious Signals

During the Miami trial, American motorists heard for
the first time such terms as "beam width," "auto-lock,"
"panning error," "shadowing error," and "batching error."
They learned that no standards existed for the beam width
of police radar, which meant the operator had no way of
knowing how wide a range of targets his unit might be
covering.

They learned that most radar units were equipped
with a feature known as auto-lock, which could be set to
sound an alarm *and lock in a reading*, at any predeter-
mined speed. As a result, an officer could dial in a number somewhat above the posted limit and go about other
business; when the alarm sounded, he merely looked
around for the handiest vehicle to which to attribute the
reading, and wrote out a ticket.

(Although some agencies had discovered that this
procedure often led to misidentification of the target, and
disabled their auto-locks before the Miami trial, the great
majority had not—and many *still* do not!)

The public learned that if the officer panned the radar
antenna across the inside of his car (for example, when
shifting his focus from oncoming to receding traffic), he
could get speed readings from his police-band or CB radio, the car's heater, ignition or air-conditioning fan, or
from the radar console itself.

They discovered that a quirk in the circuitry of some
units caused the radar to underestimate the speed of the

patrol car when it was braking hard (preparatory to turning to pursue a suspected speeder, for instance), and that it would add that error to the target reading.

They found that moving radar could sometimes mistake a nearby car, trailer or truck for the ground, and that when this "shadowing" occurred, it added several miles-per-hour to the target speed.

Most surprising of all, perhaps, was the revelation that some manufacturers had given their products names like "Speed Gun" and "Ra-Gun"—had gone so far as to equip the units with pistol grips and target sights—all in an apparent effort to deceive the police and/or public into believing that radar could be "aimed" at a particular target.

(One innocent state trooper, interviewed on national television about target identification, responded by aiming his radar down the highway and saying, "It's easy, you just point at the speeder and pull the trigger.")

Those who had been surprised to hear expert witnesses at Miami testify that moving radar might be in error as much as 25 percent of the time began wondering if the true figure might not be even higher.

There were at least three direct consequences of the Miami trial. First, the media had a field day with it; after the series on WTVJ Miami started things off, WJKW Cleveland, WAGA Atlanta, WLS Chicago and NBC's *Prime Time Saturday* all signed in with radar exposés of their own. Reporters, editorial writers and cartoonists gave extensive coverage to Emperor Radar's lack of clothing.

Secondly, after Judge Nesbitt ruled that police radar "had not met the test of reasonable scientific certainty," another Florida jurist broadened that opinion. In *Florida v. Allweiss*, Judge Karl B. Grube ruled that "absent a sufficient basis for determining accuracy, and absent a sufficient basis for confirming the competence of the radar operator, this Court finds that the [laws] . . . do not warrant the admission of radar readings as evidence in this State's courts."

Faced with a major loss of funding from radar tickets, Florida's governor moved quickly. A commission was appointed to study the issue and recommend standards for equipment and training. In recognition of the importance of his decision in *Aquilera*, Judge Nesbitt was made a member of the commission.

Thirdly, the NHTSA, faced with the embarrassment of having made radar the centerpiece of its 55 MPH enforcement program, sprang into action in an effort (a) to show that radar in its present form was "an effective enforcement tool," and (b) to find out how serious the problems really were. In February 1980, the safety administration issued a position paper (DOT/HS-805254), dispensing with any pretense of objectivity in its opening paragraph.

"The role of police traffic radar in traffic safety enforcement continues to be of critical importance," NHTSA declared, "especially in view of the safety and fuel conservation benefits of the 55 MPH speed limit. Based on tests of the six devices identified in the Dade County hearing (radar) is a reliable tool for police use when carefully installed and properly operated by skilled and knowledgeable operators."[18]

The agency was well aware that few if any "skilled and knowledgeable operators" were using radar at the time, and that no standards for training such operators existed.

A better measure of NHTSA's concern was the haste with which it began testing radar. Although claiming it had ordered the tests back in 1977, by odd coincidence when the results appeared they were for precisely the same units involved in the Miami radar trial: Kustom Signals' MR-7 and MR-9; MPH Industries' K-55; Decatur Electronics' MV-715; and CMI's Speedgun Six and Speedgun Eight.

Far from showing that the devices were "reliable tools of enforcement," the tests finally exposed the true extent and severity of the problems with traffic radar.

Guilty as charged

The tests were performed by the National Bureau of Standards (NBS) under contract to NHTSA. Among the items tested were beam width, shadowing and panning error, batching, power surge and delay for the off-on function, and several sources of internal and external interference.

Of the seven units tested,* six produced shadowing errors, adding from 4 MPH to as much as 33 MPH to the speed of targets. All but two units (the Kustom Signals' MR-7 and MR-9) picked up readings from the patrol car's AC/heater fan, and all were affected by internal CB and police-band radios. The Decatur Electronics' MV-715 produced erratic patrol car speed readings from the engine alternator, and "extreme" interference from the heater or AC fan motor.

All units showed critical amounts of lag time after use of the "kill switch," and all two-piece units produced false readings when the antenna was panned across the console. To make matters worse, most types of error occurred erratically and intermittently, meaning that even experienced operators could have trouble telling when a signal was correct and when it was being affected by interference.

---

*Two K-55s were tested, one designated for city use, the other for use on the highway, the difference being in the range specified for each.

Most damaging were the measurements of beam width. Whereas many experts believe that the maximum width allowable should be 12 degrees (in order to minimize the number of targets within the beam, hence the possibilities of error), the narrowest beam in the test group was 13.3 degrees, and the widest a sprawling 24.6 degrees.

NHTSA proceeded to draft standards calling for a maximum of 18 degrees for stationary and 15 degrees for moving radar, but did nothing to discourage the continued use of existing equipment. Under the proposed regulations, only Kustom Signals' units would be legal for use in both stationary and moving modes, plus Decatur's MV-715 in the stationary mode only. All other units now in use would in effect be inadmissible as evidence.

The coverup continues

Just as the NBS's findings confirmed the technical deficiencies of traffic radar, so the NHTSA's proposed training standard of 40 hours—24 classroom, 16 supervised field practice—underlined the lack of qualifications of most officers using radar. It was now clear beyond all dispute that the testimony at the Miami radar trial was correct. Radar was guilty as charged.

Little has happened in the years since these events took place. The proposed national standards for radar performance and radar training never were adopted. The state of Florida adopted a code based on the federal proposals, and began officer training programs, but it exempted existing radar equipment from the standard and continues to use the discredited devices.

No state government which has them has voted to repeal articles of law conferring *prima facie* status on traffic
radar.

Yet the evidence that radar does not constitute certain
proof of guilt continues to grow. In April 1985, the International Association of Chiefs of Police (IACP) published
Testing Of Police Traffic Radar Devices, a report on 24
models of the latest types of equipment. In spite of the fact
that the IACP returned some units to the manufacturers to
"make minor modifications to achieve compliance with
the model specifications," the final results recorded over
200 individual test failures, including everything from errors in tuning fork calibration to excessive sensitivity to
electromagnetic interference.[19] Typical of the problems
noted by the IACP were:

—"Four devices failed to comply during environmental tests. Three radar units did not function properly at low
temperature. One of the three also did not function properly under the high temperature test conditions, and another . . . failed during the high humidity test. The speed
display on one additional unit showed erroneous readings
during the vibration testing."

—"Three units did not meet the input-current stability requirements."

—"The permissible . . . bandwidth was initially* exceeded by three X-band and one K-band device."

—"Four radars did not meet this (antenna near-field
power density) requirement."

—"Seven radar units did not meet the (low voltage
supply) performance requirement."

—"Three units were not in compliance (with the audio Doppler requirement)."

---

*Presumably before the manufacturers were given an opportu-
nity to "fine tune" the test units.

—"Fourteen of the units did not meet the test standards for display readability . . ."

And so on, and so forth. (See IACP Report in Appendix C.)

Ultimately, radar's value as evidence will not be decided by the catalogs of its design failures and technological limitations, but rather through the deliberations of dozens, perhaps hundreds, of judges—the Nesbitts and the Grubes of this nation's courtrooms—after listening to and evaluating the testimony of people who have been wrongly or unjustly accused.

People very much like you.

Chapter Four

PREPARING A DEFENSE

Write it down, look it up, then start subpoenaing evidence.

I t has been said that "he who defends himself has a fool for a client." For a few hundred dollars, you may be able to hire the attorney who said it to defend you. If, on the other hand, you haven't got the $600–800 that is usual for this type of case—or if keeping a speeding conviction off your record isn't a matter of life or death—you may elect to conduct your own defense.

You'll be glad to know this undertaking stands a reasonable chance of success. Judges are accustomed to hearing many cases such as yours, and if you come to court well prepared, and remain polite, poised and respectful, hizzoner may go out of his or her way to help you through the ordeal. The prosecutor may even wish to avoid being inconvenienced enough to offer a reduction in charges, though there's little reason to agree unless points are dismissed also.

None of this is meant to imply that there aren't advantages in having a professional on your side. If you can afford one, an attorney is better equipped to exploit all the possibilities of getting a case dismissed or obtaining an acquittal. If you decide to go this route, look for someone with experience, and interest, in traffic cases. A group called the American Legal Network specializes in handling traffic tickets, and claims an 80 percent success rate.

They can be reached by mail at P.O. Box 1487, Baton Rouge, Louisiana, 70821, or by phone at (504) 383-2222.

If you've decided to go all out—if you're about to lose your license, and need it to earn a living; or if you're mad as hell and not going to take it any more; or if you've got the money and time and want to try your hand at writing a little judicial precedent—then you can expect to spend several thousand dollars before you're though.

The defense *pro se*

If the purposes you have in mind are a bit more modest, don't worry. Conducting your own defense (defending *pro se* in the language of the court) is what this book is all about. Reading it won't turn you into Clarence Darrow or qualify you to sit the bar exam, but it will help you take full advantage of some valuable provisions of the law: the right to represent yourself; the right to confront your accuser; the right to require the state to prove its case beyond a "reasonable doubt."

These are more than just rights, they are privileges virtually unique to English law. What takes place in the courtrooms of this nation (and few others, even in the Western world) is a true "adversary" process—an all-out battle, but one fought according to the principles of reason and justice, not power or position. Whether you win or lose the fight depends largely on the quality of the weapons you bring to it: your facts, your knowledge of the issues, and the extent of your preparations.

Ideally, those preparations should begin the moment an officer pulls you over and begins to write out a ticket. Granted, it's difficult to think clearly under such circumstances, but take note of the following points, at least, and write them down at your first opportunity:

How fast were you going?

Where did the officer come from, or where was he when he clocked you?

What other traffic was close to you?

What is your exact location (so you can find it again)?

What was the weather (rain, snow, wind, high or low temperatures all affect radar)?

Remain calm, open your window while waiting for the officer to approach, and when he asks for your license and registration, take them out of your wallet before handing them to him. Normally, he'll tell you why he stopped you ("You were traveling 75 MPH in a 65 MPH zone, sir") —so, before he begins writing you up, ask politely to be given a warning instead of a citation.

Do *not* say anything that can be taken as an admission of guilt. Say, "I thought I was driving at a safe and reasonable speed, Officer, and I'd appreciate it if you could make this a warning." Not, "Everyone else was going the same speed I was." Remember, the officer may write down anything you say on the back of the citation and use it against you in court.

Normally, when radar is used the officer will mention the fact, and may even invite you back to the patrol car to verify his reading. Don't refuse this opportunity. In fact, if he doesn't offer to show you the radar, you should ask to look at it yourself.

What you're seeking is the make and model of the unit, and if the officer is cooperative, how he used it to obtain a speed reading. In particular, you want to know whether it was used in the stationary or the moving mode, and if moving, from what direction. Some other information you'd like to have (but won't insist on if it's going to lead to a confrontation) is:

Where was I when you obtained a speed reading?

How long (over what distance) did you clock me?

How fast were you going (if moving radar was used) when you clocked me?

Does your radar have audio-Doppler, and were you using it?

Does your speed gun have auto-lock (if so, was he using it)?

Does this speed gun operate on X band or K band?

Do you have a tuning fork for the radar? How often do you use it? When did you use it last?

How old is the unit? Do many other officers use it? Do you ever have any troubles (unexplainable readings) with it?

How long have you been operating radar? How long with this particular model? This unit?

How often is this unit serviced? Calibrated?

Take note of the reading on the radar—is it the speed you are charged with? Write down the make, and, if possible, the serial number of the radar.

Obviously, you won't remember all these things—and it would take an unusually cooperative patrolman to stand around and give you all the answers. But study the list a while, and if the opportunity should arise, get as much of this information as you can.

If you've already gotten a ticket and are getting ready to fight, sit down with a notepad and pencil and try to reconstruct the scene of the arrest in as much detail as possible. Write down the answers to as many of the above questions as you can remember; and don't worry if you aren't sure what all of them mean. You'll soon know.

Returning to the scene

When an officer writes a ticket, he fills in a date, time and place where you are to appear for arraignment. In the majority of cases, you are not admitting guilt when you sign the ticket (in some jurisdictions, though, this may not be true—check to make sure, and if your signature *does* admit guilt, decline this opportunity). If your signature merely commits you to appear, fine—but be sure that you do. Otherwise, a warrant for your arrest will be issued.

In some instances, you might do well to refuse to sign even the uniform-type ticket, and instead go right to court —officer, radar and all. If you're passing through a jurisdiction, and know you won't be coming back that way, and if you have the time to spare, why not?

Normally, arraignment is about three weeks from the time of arrest, during which period there are several things to be done. Let's have an understanding; this isn't World War III, or another Holy Crusade; it's a traffic ticket; one we're going to try to beat without disrupting life's flow entirely. It will take some time and effort, but with a bit of organization there's no reason most of the steps can't be fit into your normal schedule.

For example, one of the first actions is to return to the scene of the ticket. Do this after work, or on the weekend, when you can spend about a half an hour on the site. Take along a notebook and/or sketch pad and some sharp pencils. An instant camera is not a bad idea, either.

What you're after is a complete description of the circumstances surrounding the arrest. Begin with a mental reconstruction of where you were when the officer claims to have spotted you, where the patrol car was, and how far you both traveled before he pulled you over. If you weren't allowed inside the patrol car at the time, match its position with your own car and take a look at the scene as it probably appeared to the officer. Then examine the area, sketching or photographing any significant features, paying attention to:

- Powerlines

- Highway overpasses or bridges

- Radio or microwave relay towers

- Large neon signs

- Airport or coastal radar stations

- Hospital buildings

- Airplanes taking off or landing

Any of these can produce electromagnetic interference capable of influencing police radar. If you brought a camera, take pictures of any such features you find. Remember, you must *prove* the existence of these features through photos, the testimony of the officer, or witnesses to your arrest.

Do a rough sketch of the highway in both directions (no skill as an artist is required) indicating which way you were traveling, where the officer was when he claims to have seen you, and any unusual features of terrain such as hills, curves or other obstructions.

The idea, of course, is to identify as many potential sources of false readings and causes of incorrect target information as possible with the intention of introducing them as evidence. Referring to your notes, indicate the approximate position of all other traffic on the road at the time of the arrest, with special attention to trucks, cars pulling trailers, or other large vehicles.

Don't forget to verify the speed limit in the area, noting whether it appears reasonable (except in the case of most 55 or 65 MPH highways, of course) or whether it constitutes a speed trap. Look for a sign saying "Speed limits enforced by radar," or something of that nature. In some jurisdictions, the absence of such postings is grounds for dismissal of a radar-backed ticket.

Subpoenas and depositions

A subpoena is an order issued by the court requiring a person (or object) to appear at a specified time and date. Witnesses may be subpoenaed by the prosecution or the defense, as may certain items of physical evidence.

The use of subpoenas in a simple speeding case is somewhat unusual but by no means unprecedented. In order to establish whether a radar device was operating properly, and whether it was used correctly to obtain a speed reading, considerable documentation is needed; if only one of the records or certificates necessary (in the judge's opinion) to your trial is missing, you may receive an instant dismissal of the charges.

To some extent, then, the subpoenaing of a large number of items is meant to inconvenience enforcement authorities and/or enhance the possibility that something critical will be missing or unavailable for your trial. But it also can be the means by which you are provided with information of great value to your defense.

Go to the address shown on your ticket and ask to see the Clerk of the Court. (Or you can check in your library for the proper means of issuing subpoenas in your particular jurisdiction.) If you intend to call a witness, and have any reason to feel he or she might not show up, you can provide the clerk with the person's name, address and phone number, and ask that a subpoena be issued.

Whether or not you call a witness, you definitely will subpoena documentation relating to the equipment used for your arrest and the officer's qualifications to run such equipment. The items you'll call for include, but are not limited to:

1. The radar unit itself.

2. Copies of the manufacturer's specifications for the radar, and each and all training, operator and technical manuals related to the radar.

3. Any certificates of testing and daily calibration logs for the radar.

4. Service and maintenance records for the radar.

5. The tuning fork with which the radar was calibrated, and documents attesting to its accuracy.

6. The officer's arrest records: total (past year or more) and for the day of the arrest.

7. The officer's training record and certificate to operate radar.

8. Repair records for the patrol car used.

9. Certification for the patrol car's speedometer (moving radar only).

Give the Clerk a list of these items and ask for a subpoena *duces tecum* asking to be provided with the items ahead of time for examination. Ask that the radar unit and its tuning fork(s) be brought to court on the day of your scheduled appearance.

Don't be surprised if the prosecutor counters with a "motion to protect," attempting to deny you access to the documentation mentioned above. If this happens, at your first hearing before the judge you will argue that you have need of the subpoenaed materials in order to prepare your defense. And what, exactly, is it that the police and prosecutor don't want you (and the judge) to see, anyhow?

You can say, for example, "Your Honor, I subpoenaed these items 45 days ago, the police have been totally non-responsive, and therefore I have had no opportunity to prepare my case." Follow this with a motion to dismiss the charges—which may not work, but will almost certainly get you the items you want, and a continuance to allow you time to study them.

This raises an interesting point, one which will vary from one jurisdiction to another. Sometimes the judge will allow you to enter a plea on the day of your arraignment, and if you plead "not guilty," will set a new date for the trial itself. (This almost certainly will be the procedure if you're granted a jury trial in a state where they are allowed.) If this happens, inform the judge of your subpoenas and ask that copies of the written records be provided prior to the trial in order that you may study them.

If you're in a jurisdiction that doesn't allow a jury for traffic offenses, and if the prosecutor—having been alerted to your intentions by the subpoenas you've filed—is waiting with his key witness (the arresting officer), your "not guilty" plea may be followed by a rush to judgment. About all you can do is ask for a few minutes to look through the subpoenaed documents before your case is heard—and establish a record for appellate review in case you lose. If you've been denied all or some of the materials required for an adequate defense, ask the judge for an "order to compel" the state to produce the materials. If the judge refuses and proceeds with the trial, you have good grounds for appeal.

Most of the time, you should be able to get the documents you've subpoenaed prior to your trial. Just what to look for in them—and how to use what you find—is the topic of our next chapter.

Chapter Five

CHOOSING YOUR WEAPON

Studying the evidence and deciding how best to use it in court.

The defendant has the advantage in the courtroom if he knows how to use all the tools available to him. This is as it should be. The court, after all, is a massive and generally impartial institution, charged with processing vast numbers of cases as quickly and efficiently as possible. If it does the right thing 90 percent of the time, it is one of the most reliable of all human institutions. As though to make up for its own ponderousness, it gives great latitude to defendants who are stubborn enough and determined enough to explore every possibility for winning dismissal or acquittal.

These possibilities are numerous. If, as occasionally happens, the arresting officer fails to show up on the day and hour for which your trial is scheduled, the judge may (and should) dismiss the charge. Some state, county or local laws require radar warning signs to be posted in any area where radar is used. California law allows radar to be used as evidence only where a current traffic engineering survey (study of actual traffic speeds) is available, and where the posted limit reflects the 85th percentile. You can check such matters by looking up the Vehicle Code at your local library; if these or other requirements have been violated, you may be entitled to have the charges against you dismissed.

While you're looking at the Vehicle Code (most librarians will be happy to help you find it, and even explain how to use it), take a look at the specific violation you've been charged with. Its number is written on your citation. Note that a violation is defined by several separate elements; all of these must be present, otherwise there is no violation, and there can be no conviction. The prosecution should, and probably will, seek to establish each of these elements during your trial:

1. The date, time and location of the alleged violation as stated on the citation.

2. That the defendant was the person operating the vehicle.

3. That the arresting officer was present at the time and place of the offense.

4. That it is legal to use radar at the location, e.g., the required surveys have been conducted, and the required signs posted.

5. That the radar unit has produced a speed reading.

6. That the speed attributed to the driver is in violation of the law.

As basic as these points may appear, they leave considerable room for error. Did the officer make a mistake writing down the date, time or location? Was he "on duty"? In one reported instance, a state trooper got comfortable by taking off his hat and tie and rolling up his sleeves; technically, the defendant argued, the officer was out of uniform, thus off duty. The defendant won.

The moral is, inspect each element of the citation and compare it with what you've read in the Vehicle Code to make sure you haven't been given an "easy out." If you

don't strike pay dirt, it's time to examine the documents
you've subpoenaed.

Reviewing the evidence

Whether you receive the materials you've subpoenaed ahead of time and can take them home to study, or
must make do with the time the judge gives you before
your hearing begins, the first order of business is to see if
everything you requested has been supplied.

You will be handed a sheaf of documents (possibly
far more than you need in an attempt to intimidate you),
which should include:

Related to the radar

1. Repair records.

2. Manufacturer's manual and specifications.

3. Calibration log.

Related to the tuning fork

1. Certificate of accuracy.

2. Repair and/or calibration record.

Related to the officer

1. The officer's arrest records (both long term and for the
    day of your arrest).

2. Officer's radar training record and certificate for operat-
    ing radar.

Related to the patrol car

1. Speedometer calibration certificate.

2. Speedometer repair and maintenance record.

3. Patrol car's repair record.

If anything you've asked for is missing, make a note of
it.

The radar unit

As for the radar unit, it's unlikely you'll be allowed to
take this home to study. So, the first good look you'll get at
it (other than when you saw it in the patrol car) is in court.
Is it, in fact, the same unit used for your arrest? If the make,
model or serial number doesn't jibe with your notes, or
are the wrong ones for the records you've been given, the
prosecution's case may be in trouble.

Next, let's have a look at the repair log and calibration certificate. The certificate is issued by the manufacturer, affirming that the unit was working when it left the
factory (it really isn't too important, though it should be
there). The log matters more because it purports to show
that a daily test and calibration of the unit has been made
in accordance with established procedures and the manufacturer's recommendations.

Some police departments keep their logs haphaz-
ardly, or not at all, due to the paperwork involved. Such a
lapse can be helpful to your case.

Has the radar been in and out of the shop several
times in the past few days, weeks, months? If so, you can
argue that this unit is erratic and unreliable: how can the
court be sure it was working properly on the day you were
arrested? Conversely, if it has seldom or never been re-
paired, this could be a sign of neglect.

The tuning fork

As with the radar, you won't get to see the tuning fork until you get to court. Check it for essentially the same things you did the radar: is it the proper unit for this radar? Has a certificate of calibration been provided? Without such a certificate, the performance of the fork is not "traceable to a standard," making it inadmissible in some states by law. The same argument may even succeed in states without such a law.

So far as the forks themselves are concerned, they are pretty rugged; unless one has been badly nicked, bent or otherwise damaged, there's not much reason to challenge it. Don't worry, though, in this game you only need to score once to win—and there will be plenty of opportunities.

The police officer

Let's go over the officer's arrest records now. Depending on how these are kept, you may be able to tell what kind of cars he likes to cite: sports cars, hot rods, old cars or whatever. If there is a clear pattern of prejudice against the kind of car you drive, make a note to bring this out when you get the officer on the stand.

Another possibility is that the officer favors certain locations and spends an unusual amount of time at them. If a very high percentage of his tickets are written in just a few locations—and the place where you got yours is one of them—you may have uncovered a "cherry patch," a spot where many violations occur because of poor engineering, bad signing, faulty traffic controls or the like. Another term for this sort of situation is "speed trap," and most judges object to them.

If you have enough time, it could pay to return to the patch where you were plucked and check with some of the local merchants or residents to see if the police are known to write a lot of tickets there. If so, you might want to go back with a camera to make a record of the proceedings for later use as evidence. If Officer Such-and-Such habitually writes a high percentage of his citations in one location, this may be all the proof you need that he is fishing from a barrel.

When you're finished with the arrest records, take a look at the officer's radar training. Has he had 24 hours in the classroom and 16 hours of supervised field practice as recommended by NHTSA? A mere certificate stating that he is "qualified" may not mean much; one radar manufacturer, MPH Industries, is known to have provided such certificates in quantity without requiring any proof of actual training whatsoever. If the officer cannot testify to a sufficient amount of initial training, plus some retraining in the past five years—plus training in the specific model he now is operating—his shortcomings may work in your favor.

The patrol car

Examine the records for the patrol car. Does it by happy circumstance have a history of electrical problems? This could be significant if the radar is operated from the car's power system, as usually is the case. Equally important, if you were cited by moving radar, do calibration logs indicate problems with the car's speedometer? It's essential that the officer be able to compare the patrol car speed reading on his radar against the speedometer to guard against shadowing error and other forms of interference; he can't perform this task if the speedometer isn't accurate.

There should be proof that the speedometer has been calibrated recently and is working properly: a certificate looking something like the one in Figure 3.

FIGURE 3
Sample Speedometer Calibration Chart

DEPARTMENT OF CALIFORNIA HIGHWAY PATROL
SPEEDOMETER CALIBRATION CHART

DATE				VEHICLE NUMBER				
MILEAGE				TEST EQUIPMENT SERIAL NUMBER				
NAME(S) OF PERSON(S) MAKING CHECK								

VEHICLE READS	30	40	50	55	60	65	70	80	90
ACTUAL SPEED									

CHP 227 (REV 10-74)   USE PREVIOUS EDITION UNTIL DEPLETED

Adding it all up

Now that you've examined most of the physical evidence, it's time to draw some initial conclusions. What have you discovered so far that might be of help in your defense, and what is the best way to make use of it? Depending on the exact circumstances, there really are only three possibilities:

1. You can seek to prove that the speed reading itself was incorrect;

2. You can seek to prove that, while the reading may have been correct, it was caused by a vehicle or object other than your car; or

3. You can show that, while the speed reading may have
    been correct, and may have been caused by your vehi-
    cle, it was obtained through the use of a speed trap and
    should be disallowed.

The final option is the weakest since it involves a tacit
admission of guilt. If you were out there all by yourself,
and there were no likely sources of electrical interference
in the neighborhood, and both the officer and the radar
unit appear competent, then you may have no other
choice. But this is a defense to be used only *in extremis,* as
a last resort.

In most situations it will be easier to prove error in, or
reasonable doubt about, the speed reading or target identification. In developing your defense, you're going to call
the court's attention to:

1. The presence of all possible sources of interference in
    the area.

2. Other traffic which may have been responsible for the
    speed reading.

3. Any deficiencies you've discovered in the radar, the of-
    ficer's training, his methods of operating the radar, or
    with the patrol car.

4. Any technical violations you have found in the require-
    ments for the use of traffic radar in this jurisdiction.

The point of all this, in case you're wondering, is to
attack what the court refers to as the Due Process Connection. According to the Due Process Connection, in order
to sustain a conviction there must be a *necessary connection,* rooted in common knowledge, between the fact observed (the radar reading) and the fact presumed (your
speed). Throughout your trial, one way or another, you

will be attacking this "necessary connection" by showing that police radar can and does produce readings in the total absence of a valid target, and that no necessary connection between the reading produced and your vehicle exists.

Perhaps a couple of examples would be instructive:

> You were driving along a divided urban Interstate when you were stopped and cited for going 70 MPH. Conditions were: daylight, fair weather, moderate traffic, with power lines and a chain link fence parallel to the highway. The radar, a "Pulsemaster MLV," was used in the moving mode.

In court, you will attempt to show that there were several possible sources of interference present (the power lines and the fence); that there was sufficient other traffic, including large trucks, to make target identification questionable; and that the beam width and range of this particular radar make it more susceptible to target error than some other units. In short, that there is reasonable doubt the radar's reading was correct, or if it was correct, that your vehicle was the one responsible.

Another example:

> After rounding a curve on a four-lane, divided highway, you were stopped and cited for going 48 MPH in a 30 MPH zone. Conditions were: daytime, fair weather, light traffic, with some traffic on a frontage road parallel to the highway. Instant-on radar—K-band "Duckshooter"—was being operated in the stationary mode from a patrol car parked on the median.

In court, you will attempt to show that traffic on the frontage road—directly in the radar's beam, but moving *away* from it—was equally capable of producing the radar's reading; that the absence of a "tracking history" (because the radar was used in the instant-on mode) made it impossible for the officer to tell which vehicle was responsible for the reading; and that the highway is not posted with radar warnings (if it wasn't) in accordance with the law, possibly making the state's evidence inadmissable.

The potential for developing a good defense is almost unlimited. Where limits do apply is in the order in which you present your case, the kinds of evidence that are admissible, and the means you may or may not use to introduce it.

Chapter Six

CHOOSING YOUR GROUND

Arraignment and entering a plea; some initial maneuvering; how to select a jury.

Getting off to a good start is important, especially in legal matters. This is not the time to be casual. The time, date and place of your first appearance in court should be on the front of your citation. If for any reason you can't show up, you must advise the court or have your attorney appear in your place to ask for a continuance. Failure to do one or the other will cause an arrest warrant to be issued in your honor.

If at the last minute you fall ill, call the Clerk of the Court and tell him your problem. As soon as you are able, follow this up with a registered letter referencing your phone call, giving your reason for not appearing, and asking for a new court date. Remember, if a warrant *is* issued, you'll either be picked up at your home or arrested on the spot the next time you're stopped for any reason. So don't slip up.

In some jurisdictions, the date on the ticket is for **arraignment**. At this time you'll be informed of the charges against you, advised of your rights, and asked to enter a plea. If you plead "not guilty" and request time to prepare your defense, a continuance almost always will be granted. You are legally entitled to a proper defense.

Other jurisdictions, in particular those that do not allow jury trials for traffic offenses, consider the time between the issuance of the citation and the date of appearance as adequate for preparing a defense. In this event, you may be appearing for **arraignment and trial**, and be required to present your defense on the spot. If it isn't clear from the wording on the ticket whether you are to appear for arraignment only, or for arraignment and trial, call the Clerk of the Court and ask for an explanation.

Keep in mind that the judge has almost unlimited power in his own courtroom. He may grant a continuance or not, regardless of what the Clerk has told you; may decide to proceed with the case because the arresting officer is there to testify, or dismiss the charges if he isn't. It's all up to him.

The best plan, assuming you've had two or three weeks to issue your subpoenas and otherwise prepare, is to go into court ready to put on a defense. That way, you won't be caught short.

States that allow you to plead "not guilty" by mail usually have instructions for doing so printed on the citation. Generally, it's required that you write a letter waiving your right to a formal arraignment, entering a plea, and stipulating whether you want to be tried by a jury or in front of a judge. Some jurisdictions also require you to post an appearance bond, usually about $25.

Prior to your arraignment, call the Clerk of the Court and find out what the usual fine for your violation is and whether the court accepts personal checks, money orders or auto-club membership cards. This way, you'll be ready if the judge asks you to post an appearance bond in the amount of the fine.*

Although the exact procedures vary between states, the purposes of arraignment are the same everywhere: to

---

*Technically, your appearance for arraignment shows good faith and should entitle you to release on your own recognizance— but traffic courts tend to ignore such niceties.

advise the defendant of the charges against him and explain his options. Arraignment also is the time to enter a plea or to request a continuance.

There are four types of pleas, "guilty," "guilty with an explanation," "_nolo contendere_," (no contest) and "not guilty." Guilty with an explanation probably is heard most often, being a way of allowing the defendant to vent some of his frustrations, and the judge to get on with assessing a fine—perhaps slightly reduced in honor of a good story.

Nolo contendere is how you plead guilty in Latin.

Obviously, the fourth choice is the only one that interests you, or you'd have mailed in your fine to begin with.

"Not guilty, Your Honor"

The arraignment probably will cost you half a day—a morning or an afternoon—depending on where your name comes on the list of cases to be heard. Dress as neatly and conservatively for this occasion as you did for your Confirmation or Bar Mitzvah: clothes may not make the man, but they most certainly make the first impression, which is important in court.

When you arrive, find your name on the court docket —a computer printout posted in some conspicuous place —telling you where your hearing is to be held. Once there, identify yourself to the bailiff or Clerk of the Court— then have a seat and prepare to watch democracy in action. If you're called early, breathe a little prayer of thanks; if not, take advantage of the opportunity to study the system.

When your name is called, step to the microphone in front of the bench. The judge will read the charge and ask how you plead. After pleading not guilty, you'll be given the option of a jury trial, *if* your state allows it.

Assuming you are permitted to have a jury, you must decide whether or not you really want one. Studies of traffic cases show that judges actually may be more sympathetic to the defendant than juries. Jury trials require more time both for preparation and presentation, and may evoke a stronger effort by the prosecution to convict. In short, jury trials tend to raise the stakes for everyone involved.

If you opt to forego a jury, the judge normally will set bail and order you to appear on a certain date to stand trial. Resist any invitation to plead guilty with an excuse. Also decline if the judge invites you to "tell me all about it," as this is tantamount to waiving your right to trial. Merely reply, "I don't have anything to say until I hear the State's case."

Ideally, you hope the trial will be delayed for a while, long enough that the officer's memory of your arrest will fade; that he will be transferred, move out of town or resign; that your records will be misplaced or destroyed, or that some other nice thing will happen. It makes sense, therefore, to ask for as long a continuance as possible (the usual maximum is 90 days), but be prepared to explain why you need the extra time to prepare your defense.

At the same time, the Constitution guarantees you the right to a "speedy" trial, and if the time limit specified by the law in your state is exceeded, you're entitled to dismissal. Reply to any other "suggestions" by the judge by saying, "Your honor, I want to see the State's case before I do anything." And if you're asked to "waive time"—extend the period the state has in which to try you—you will decline politely but firmly. Remember, asking for a continuance may waive your right to a speedy trial.

Assuming a date has been set, and it's within the proper time limits, your next step is to check up on your rights of "discovery." You're entitled to see all of the prosecution's evidence prior to your trial—that means all the items you've subpoenaed. If the police or the prosecutor try to hold back, hasten to the library again to check out the Rules of Discovery in the laws of your state.

The prosecutor may be up to trying some other maneuvers. If shortly before the trial he contacts you for a session of "Let's Make a Deal," be careful. Occasionally, his objective is simply to save time and effort; but it can also mean that his case is weak or that he isn't ready to go to trial, in which case a deal may not be to your advantage. Even if he is prepared to have your fine suspended, the violation will appear on your record, possibly resulting in higher insurance premiums, plus increased fines in case of future violations.

Unless you get everything you want, which means complete acquittal, there's little the prosecutor can offer of any real interest.

Be prepared to respond with a resounding "no" if the prosecution offers you a trial "by deposition." In a trial by deposition, the evidence against you is the arresting officer's sworn statement of the "facts," which means you have nothing to cross-examine but a piece of paper. The Constitution guarantees the defendant the right to be confronted by his accusers; don't bargain away this right.

Should you testify?

An important element of pre-trial strategy is your decision whether to take the stand in your own defense. The first thing to realize is that you are under no obligation to testify. A traffic ticket is not a matter of national security, real or imagined, and unless the prosecution calls you as a witness (highly unlikely), you won't even have to plead the

Fifth Amendment. You will simply limit yourself to making motions, cross-examining the prosecution's witness,
and summing up—and you will bypass both direct testimony and cross-examination by the prosecutor. Remember that uncontroverted testimony stands as proof; you
must usually provide or elicit testimony denying the key
assertions by the state.

If you're defending yourself in front of a jury, it could
be a good idea to testify. It will help satisfy the jurors' curi-
osity about your side of the case, and if the prosecutor gets
too aggressive with you, you may win sympathy. If, on the
other hand, you are short tempered, easily confused or so
shifty-eyed you can't meet your own gaze in the mirror,
you may be better off passing. In fact, you might be better
off not having a jury.

By the same token, if you are guilty as charged, but
believe that the speed limit was improperly derived and
prejudicially applied, it would be unwise to take the
stand. You don't want to be in the position of either having
to confess, or having to lie under oath, since such acts can
be dangerous to your wallet, your freedom, or both.

Jury selection

The final preparation is the selection of jurors. This is
something of a specialized process, and many lawyers set
great store by their ability to single out persons who will
be well disposed toward their clients. You may do almost
as well as the lawyer by remembering that males age 25 to
45 are the most representative of "average" motorists, and
may also be more sympathetic to persons accused of vio-
lating unreasonable speed limits.

Prospective jurors are drawn at random from the rolls
of voters, meaning you should have a fairly wide cross sec-

tion to choose from. Some jurisdictions empanel 12-member juries, other may have as few as six; regardless of the number to be chosen, the process will begin with a few words from the judge instructing you on your rights and responsibilities in the matter of choosing the jury.

The prosecutor begins *voir dire*, which means questioning the prospects about their jobs, their beliefs, whether they have ever received a radar-based speeding ticket, etc. He will try to choose jurors he thinks will be more willing to return a guilty verdict; you should make notes on the responses he gets and refer to them while making your own choices.

When your turn comes, ask each prospect if he or she understands that, until the prosecution proves each element of the charge beyond a reasonable doubt, you are not guilty. Try to find out if the prospect would agree that police radar can malfunction under certain circumstances, and that it is possible for a police officer to make mistakes using radar. Note the answers and any gut reactions you have about the person you're questioning before reaching a decision.

There are two means of rejecting an unsuitable prospect. You can challenge "for cause" when you have reason to question the person's ability to be impartial (such as the family doctor you just sued for malpractice), or you can employ several "preemptory challenges," which do not require you to give any specific reason. Don't feel obligated to use all of the preemptory challenges you're allowed; unreasonable challenges waste time and will not gain you points with the judge or other jurors.

Eventually, the requisite number of jurors will be accepted by you and the prosecutor, and as soon as they are impanelled, your trial is ready to begin.

Chapter Seven

YOUR DAY IN COURT

How the game is played; some "do's" and "don'ts" for the defense.

There is something fundamentally unsettling about hearing a stranger's voice call out: "State, People, or Commonwealth versus—**your name.**" The odds seem so unfair. In reality, though, there's little to worry about if you've done your homework properly.

In some respects, you even have the advantage. You've had more time to study your case than the prosecution—which must worry about many other cases coming to trial. You know which parts of the prosecution's case are most vulnerable to attack, and the line of defense you're going to pursue, while the prosecution can only guess about these things.

In fact, the prosecutor's only edge is that he may understand the mechanics of the trial—the rules and sequence of events—a little better than you do. And we're about to remedy that.

The players and program

After the judge announces your case, and asks the defense and the prosecution if both are ready to proceed, there is an opportunity for opening statements. The prosecutor goes first, and will state that he intends to prove the following elements: that you, the defendant, were observed by the arresting officer at a certain time and date in a certain place; that your speed was recorded by a properly calibrated and functioning radar; that a citation was then issued in accordance with the law; and that he, the prosecutor, will show that you, the defendant, are guilty as charged. (If he leaves out any of these, it may provide you with an important clue about where to strike.)

You may be allowed to respond with an opening statement if you wish, but there's little need to do so. In cases tried without a jury, opening statements usually are dispensed with entirely (the judge has heard it all before and, in theory at least, isn't impressed). Even when a jury is present, the defendant may do well to forego an opening statement, saving his main points until they come out on the stand or during summation. If called on, merely reply, "The defense waives its opening statement, Your Honor."

The prosecution now begins to present its case. It calls the arresting officer, who is duly sworn in and takes the stand. The prosecutor establishes the witness's credentials and competence to testify, then questions him in an effort to bring out the state's or commonwealth's version of what happened. This phase of the proceedings is known as "direct examination."

During direct examination, only simple questions of fact may be asked or answered. Questions that suggest an answer, or way of answering, are known as "leading questions" and may not be asked during direct examination.

For example, the prosecutor should ask the officer, "Did you observe a vehicle? Were you able to determine the speed of the vehicle? Was the vehicle's speed in excess of the limit?" He may not ask, *"Did you see the defendant speeding?"* because this calls for several conclusions (was it the defendant's vehicle, was the speed in excess of the posted limit, was the defendant driving the vehicle, etc.) that have not yet been established.

If the prosecutor does appear to be leading the officer's testimony, you have the right to object to both the question and any answer that has been given. Rise and politely object to "the introduction of facts not in evidence." The judge will either sustain or overrule.

As the prosecutor proceeds with his questioning, he will try to bring out all the points necessary to establish guilt. In a civil case being tried before a judge, the standard for proving guilt is "a fair preponderance of the credible evidence," whereas in criminal cases argued before a jury, it's necessary to establish guilt "beyond a reasonable doubt." Although this suggests that guilt requires less proof in civil cases than in criminal ones, as a practical matter "reasonable doubt" remains the accepted criterion in most jurisdictions for finding a defendant innocent.

Every court operates under generally similar rules of evidence. To be admissible, evidence must be **relevant**—that is, must bear a relationship to the fact to be proved. Evidence must be **competent**—that is, must be supported by proper authority. (This book, for instance, is not admissible as evidence—because you, the defendant, didn't write it, and the authors aren't present to testify to its accuracy.) Evidence must be **material**—that is, must be of significance to the case. If, for example, stationary radar was used to obtain a speed reading, the fact that the patrol car's speedometer hadn't been calibrated recently would be "immaterial."

The prosecutor may ask the officer to state what happened in his own words, or may take him through his testimony point by point. In either case, help may have been received from the radar manufacturers, who often supply "canned testimony"—standard statements, or format question-and-answer lists—to be used against defendants.

By memorizing formula testimony, an operator is encouraged to recall what *should* have happened, regardless of what actually did happen. He receives instructions on how to describe the operation of his radar unit, his certificate of competence, the reliability of the Doppler Principle, and how in the present case—with the speed meter operating in a normal manner, and the calibration checked before and after by a certified tuning fork—he "observed the defendant exceeding the posted speed limit, and obtained stable reading before issuing a citation."

Canned testimony is designed to stand alone, and quite often comprises the whole of the prosecution's case. Historically, such a small percentage of motorists has challenged radar, and has had such poor success doing it, that nothing more was necessary. Why tinker with a winning formula?

Let's see if we can give them some reasons.

What's in the can

Unless each element of your alleged violation is proved beyond a reasonable doubt, you can move for dismissal the minute the prosecution rests its case.*

Exactly how many elements must be proved, and what they are, depends on local laws—which is why you took the trouble to look them up in the library. Keep a checklist and note which points have been covered; if the prosecution omits, or fails to establish, even one of these items, you may have hit pay dirt.

---

*Make sure the prosecutor has rested; if he doesn't say so in so many words, query the judge about it.

Here is a typical example of canned narrative testimony:

> My name is John J. Smith. I am a patrolman assigned to the traffic enforcement division of the Johnson City Police Department. As part of my duties for the past year, I have operated police radar for this department. I hold a certificate of competency issued by the manufacturer's radar training instructor, which is dated October 8, 19–.
>
> On the afternoon of December 3, 19–, I was operating a radar speed check on South Main Street at the corner of East Seventh Avenue. The radar speed meter in use was a Doppler radar speed meter manufactured by Kustom Signals, Inc. of Chanute, Kansas. The radar speed meter was operating in a normal manner. Calibration had been checked when the radar speed meter was set up, at 2:15 P.M. and again at 3:20 p.m. by use of a certified tuning fork.
>
> At approximately 4:25 P.M., a 19– Pontiac sedan, Kansas registration number NO-1750, later found to be driven by the defendant, J.O. Jones, was observed approaching from the south at what appeared to be a high rate of speed, which speed, from visual observation, appeared to be some 50 miles per hour. At the corner of East Fifth Avenue and South Main Street, the defendant's vehicle was out front, by itself, nearest the radar speed meter. At that point, a stable radar speed meter reading was obtained on the vehicle operated by defendant Jones. The speed reading thus obtained was 55 MPH. The posted speed limit on South Main Street at the intersection of Fifth Avenue is 35 miles per hour.

Once the officer has made his statement, he has little more to do than point out the defendant, and the prosecution usually can rest its case.

Let's take this testimony apart and see how well it covers the alleged offence.

First of all, it establishes nine primary elements necessary for conviction. The **driver** (J.O. Jones) of the **vehicle** (19– Pontiac, license NO-1750) was **speeding** (55 MPH) in a zone posted for a **speed limit** of (35 MPH). The violation took place at a specific **time** (4:25 P.M.) on a specific **date** (December 3, 19–) at a specific **place** (corner of East Fifth and South Main) where a **police officer** (on duty) was operating a (properly calibrated) **radar** speed-measuring device.

Secondly, note that this testimony relies on the decision in **Honeycutt,** that the vehicle "out front alone, closest to the radar," is responsible for its reading. If allowed to go uncontested, this is all that's needed to gain a conviction.

If the leading actor in this little drama isn't well enough rehearsed, the prosecutor may take him through his performance question by question. Here's an example of "Q & A" testimony from one manufacturer's Operator's Manual:

Q: *"How much experience do you have operating police radar?"*
A: (Answer specifically how long.)

Q: *"On the date in question, what type of radar were you operating?"*
A: "Doppler radar speed meter (model) manufactured by (company) of (city address)."

Q: *"How do you know it was a Doppler radar?"*
A: "It is so marked on the label, in the instruction manual, and by instruction of the training representative."

Q: *"Have you any formal training in this particular device?"*

A: "Yes. The manufacturer conducts a course of classroom and street training which I attended. Here is a certificate of competency issued by the manufacturer."

Q: *"Did you check the calibration of the radar?"*
A: "Yes, at (time)."

Q: *"By what method?"*
A: "By the use of a serialized tuning fork furnished by the manufacturer. I strike this fork on my shoe heel and hold it in front of the antenna. The radar is supposed to read 50 miles per hour when this is done."

Q: *"How do you know the tuning fork is accurate?"*
A: "Each fork is accompanied by a certificate of accuracy. Here is the certificate on the fork in question."

Q: *"Was the radar in calibration according to that test?"*
A: "Yes, sir."

Q: *"What do you do if it is not accurate?"*
A: "I shut the machine down and turn it in. There are no external accuracy adjustments."

Q: *"Did it appear to be operating normally?"*
A: "Yes, sir. I had the machine in operation for approximately (hours, minutes) and it was reading each vehicle that came through with steady readings. It was not showing any abnormal readings or characteristics."

Q: *"Where were you operating the device?"*
A: (State the location.)

Q: *"Where did you first see the defendant's vehicle?"*
A: (State the approximate distance he was first observed and the direction of travel.)

Q: *"What attracted your attention to the defendant's vehicle?"*

A: (State whether he was going at what you judged to be excessive speed. State if he was passing, overtaking, or drawing away from other traffic.)

Q: *"What did you do then?"*
A: "I aimed (or had the radar aimed) in the direction of the defendant's vehicle and noted the reading."

Q: *"Was the reading stable and consistent with what you visually judged his speed to be?"*
A: "Yes, he just seemed to be going quite a bit faster than the speed limit, at which most of the traffic was traveling. I watched the reading for about half a block and it was stable at (MPH). I have been instructed to and always relate my judgment of the vehicle speed to the reading obtained by the radar."

Q: *"Was his the nearest vehicle to the machine?"*
A: "Yes. He was out front, by himself nearest the radar."

Q: *"Did you lock the reading on the radar device?"*
A: "Yes."

Q: *"What was the reading on the radar device?"*
A: (State the speed.)

Q: *"Did you offer to let the defendant look at the reading?"*
A: "Yes, he did (or refused)."

Q: *"In the case of a violation where your vehicle was moving, how do you check the computation?"*
A: "I go through the sequence of first recognizing the vehicle as a possible violator, I then check that he is the vehicle out front nearest the radar, if the reading is consistent with what I judge his speed to be I lock all the readings. The radar has a separate set of readings which I check against my calibrated speedometer at the same time

I lock the readings. This assures that the radar computation is correct."

Apart from being disingenuous enough to make a hyena gag, the above testimony covers a couple of points that the narrative example missed. Here, the officer affirms, and reaffirms, that he observed the suspect speeding *before* making use of the radar to verify that impression ("I have been instructed to and always relate my judgment of the vehicle speed to the reading obtained by the radar"). Which is the recommended, and almost universally disregarded, procedure.

Whether the prosecution's testimony is canned or not, pay attention to the exact wording by which each of the main elements of the case is covered, and whether it is covered. If the officer and prosecutor leave out an element, make a note of this fact, but do not refer to the point during cross-examination. Instead, after you've finished cross-examination, and the prosecution rests (make sure the prosecutor says, "The state rests, Your Honor," or whatever), move for dismissal based on the fact that "the prosecution has failed to prove such-and-such an element of the crime." If the prosecutor *then* tries to remedy his error, you will object, reminding the judge that "the prosecution has rested and is prohibited by law from re-opening its case." If it is allowed to do so, you have made a good record for appeal.

Your turn at bat

When the prosecutor has finished *direct* examination, it's your turn to "cross-examine." This is the most important part of your case, and your best chance to create reasonable doubt about the circumstances pertaining to your arrest.

Even if you've decided to testify later, your version of what happened will carry less weight with the judge and/or jury than an admission, implicit or explicit, by the officer that the speed reading he obtained may not have been correct or may not have been caused by your car. Cross-examination is covered in the chapter just ahead.

Chapter Eight

THE OPENING ROUND

Direct examination by the prosecution and cross-examination by the defense.

Questioning the officer who arrested you may feel uncomfortable at first, but you'll get used to it. You'd better. It's vital to a successful defense. Only through cross-examination can the facts essential to your acquittal be brought out and made a matter of record.

If you're innocent of exceeding the speed limit, there must be some explanation for the officer's error—and this is the procedure through which you (and the court) will discover what it is.

During this phase you can only ask questions and clarify answers. You cannot comment on those answers or otherwise give testimony yourself. The temptation may be overwhelming, but if you give in to it, you incur the wrath of the prosecutor and the judge.

Your goal throughout cross-examination is to smooth the way for the officer to give answers that support *your hypothesis* of what happened, or force him to admit uncertainty about *his* version. Both strategies are based on exposing the most frequent failures of radar, which are:

1. Failure to identify the proper target.

2. Failure to obtain a correct speed reading for the target.

3. Failure to detect and avoid spurious speed readings from non-target sources.

4. Malfeasance and/or incompetence on the part of the officer.

If during cross-examination you can establish the possibility or probability that one or more of these failures occurred, you'll be well on your way to winning acquittal.

Examining the physical evidence

In addition to these essentially *operational* errors, there are possibilities for technical or procedural errors as well. So far as conviction or acquittal is concerned, the court makes no distinction between the two, and neither should you. There is nothing wrong with holding the officer and his department to strict rules of procedure, and seeking dismissal if such requirements are not fully met.

For this purpose, a careful review of the "physical evidence"—the materials you subpoenaed—can be important. If any of this material reveals that the technical requirements of the law, the local jurisdiction or the enforcement agency have not been met, it will be your business to make the court aware of the fact at the proper time.

As soon as the prosecutor finishes examining the officer, and concludes with, "No further questions, Your Honor," you will be called on to begin cross-examination. You may be allowed to do this from your seat at the defense table, or may have to go to a microphone facing the judge's bench and witness stand. By the time your case is called, you'll probably know which it is, but if in doubt simply ask the judge.

Unless you're blessed with total recall, you should make a list of questions to use during cross-examination; adding to it as necessary, and making notes on the answers you get.

Presumably, you've examined the documents you subpoenaed (see list on page 46). If any of these items has not been provided, you will make a note to ask the officer why your subpoenas were not honored fully. (This makes his answer a matter of record.)

You will have checked the dates of the documents to see that they're current, and made a note of any that aren't; examined the officer's arrest records to see if (a) he spends a lot of time writing tickets in the same location, or (b) wrote several tickets in addition to yours for the identical speed. (This last would suggest that he locked in a speed reading and used it several times.) If you've found evidence of any of this, add questions to establish these facts during cross-examination. Refer to the answers when the time comes to move for dismissal or during your summation.

Look at the calibration log to see how often and at what times the radar was checked for accuracy. Two court decisions, _Wisconsin v. Hanson,_ and _Minnesota v. Gerdes,_ established that a calibration check with a tuning fork should be performed "within a reasonable time" after a citation is issued.

In two other cases, _Connecticut v. Tomanelli_ and _New York v. Stuck,_ tuning fork tests were conducted immediately before and after the citations were issued (_Stuck_ specifies that tests took place at 11:15 P.M. and 11:45 P.M., establishing reasonable proof that the radar was functioning properly at 11:35 P.M.). All of these cases indicate that merely calibrating the radar at the beginning and end of an officer's shift—as is often done—is *not* adequate to guarantee the radar's function at the time of the arrest.

When you've finished looking at the documents, ask to see the radar unit and the tuning fork. If either is missing, ask the officer why; his answer may supply a reason for dismissal or acquittal, or may form the basis for an appeal if the decision goes against you.

If both items are present, check to verify that their serial numbers match those in the documents you've been given; if not, call the fact to the judge's attention at the conclusion of your case and move for dismissal.

If your research supports your position, you should object to a particular radar unit as soon as it is identified in the officer's testimony. "Objection, Your Honor, this particular device has been denied (or not accorded) judicial notice in this appellate jurisdiction, nor does it meet the federal minimum performance standards. Any readings from it should not be admitted without expert testimony."

Be prepared for one of three responses by the judge:

1. "Well, it's admissible in my court, overruled." (Excellent grounds for appeal.)

2. If judicial notice has been specifically denied, the judge should ask for the case. If you can supply a copy, the trial could end.

3. If judicial notice has not been accorded to the unit, the burden falls on the state. The prosecutor must establish the admissibility of the evidence (though you may have to remind the judge of this fact). The prosecutor may cite one or more cases admitting radar. Your task is to show that they do not apply to this particular device. Researching the laws in your jurisdiction and referring to the case histories in Appendix A of this book may provide the information you need. If so, all or part of the following statement is appropriate:

"Your Honor, the case(s) that the prosecutor cites do not involve this jurisdiction. In addition, they do not refer to this particular device. This case took place (N number) of years before this device was even developed and cannot possibly be applicable here."

Many judges hesitate to rule on such questions, and will "take them under advisement." This effectively suspends the trial, usually for good. If the judge rules in your favor, you win. If he rules against you, you have good grounds for an appeal.

If you're fortunate enough to find something really promising in the physical evidence, close in on it like a mongoose on a cobra. First, ask the officer to identify just those documents you are interested in; when he has done so, offer them to the judge as evidence (don't worry if you don't know the legal terminology for this sort of thing, just say what you want plainly and politely).

If the prosecutor objects to the introduction of a document or a particular line of questioning (rather likely), don't be rattled. Instead:

1. Go ahead and finish your sentence.

2. Stop and listen politely to the objection.

3. Reply with, "Your Honor, my question is relevant because..." and explain the importance of the document or question to your defense.

Cross-examining the officer

You're ready now to begin examining the witness in earnest. You have prepared a set of questions designed to establish your hypothesis—your version of what actually happened—and to draw the officer out on possible ways he could have got the wrong speed reading or the wrong target.

Before beginning work on your hypothesis, you should cover any questions about the adequacy of the officer's radar training, and/or discrepancies you have uncovered in the subpoenaed evidence. It's good technique to start with a few easy questions to put the witness at ease and get him into a pattern, or rhythm, of familiar responses. That way, when you get to the question that counts, he may not think too hard before answering. Some questions to start with are:

"Officer, what make and model radar were you using at the time of the alleged violation?"
The answer should agree with the facts, of course.

"Is this unit designed to be used in the stationary or the moving mode?"
It may be one or the other or both.

"Which mode was it in at the time of the alleged violation?"
This should be enough to get things rolling and set up your next group of questions.

If the officer's training in radar was covered during direct examination, and if it met or exceeded NHTSA's recommended 24 hours classroom, and 16 hours supervised field experience, you can skip this set of questions. Or, if you begin and find that training has been adequate, simply move on to your next line of examination.

On the other hand, if after the first couple of answers it looks like you've discovered a weakness, bore in with the following:

"Officer, with regard to your formal training in the use of radar, would you tell the court where you received that training, and who examined you?"

If he went to a special school, good for him; if he got it at the station house, that may be good for you.

"From whom did you receive your training?"
If from a manufacturer's representative—or a fellow officer who got *his* training from the rep—you may be nearing pay dirt.

"How many officers were in the group you trained with?"
Just a general sort of question: if he was alone, the training may have been pretty informal; if with a large group, he may not have received enough individual attention.

"What did your training consist of—how many hours, and with what types of radar?"
You're hoping it was less than the recommended "24-16," and with some radar other than the make and model used for your arrest. If so, you'll follow up with:

"Are you aware that many agencies require a minimum of 24 hours of classroom and 16 hours of supervised field experience for training in the use of radar?"
If the judge asks you what your authority is, or which agencies have such requirements, tell him that the information is contained in "Police Traffic Radar," published by the National Highway Traffic Safety Administration, February 1980, reference number DOT HS-805 254, available from the Department of Transportation. Add that most modern police departments have adopted NHTSA's standards, including all agencies in such states as Florida and Michigan, and that the officer should be aware of them from his radar training course. The prosecutor may succeed in having all of this stricken from the record as "incompetent," but you will probably have made your point.

"Inasmuch as you did not receive your initial training with the (name of radar unit used for your arrest), can you tell the court how much training you have received in the use of that particular unit?"

With luck, the answer may be little or no special training.

"How long ago did you receive your initial training, and how long has it been since you attended a refresher course in the use of radar?"

NHTSA calls for retraining every one to three years.

Remember, carry this particular line of questioning only so far as to cast doubt on the officer's expertise; if you find you are not succeeding—or as soon as it becomes apparent that you have succeeded—it's time to move to your next set of questions.

Employing the physical evidence

Let's suppose you've had a bit of luck and found something in the subpoenaed materials that could be helpful to your case. It might be any of the things we've discussed, but for the sake of illustration, imagine that you've looked at the calibration log and found that the radar was *not* checked with a tuning fork in the several hours before or after your arrest. You can establish this fact during cross-examination with questions like these:

"Officer, have you seen this document before?" (thus authenticating, or "proving up," the document obtained by your subpoena).

Yes.

"Are you familiar with the term 'calibration' as it applies to radar?"

Probable answer is yes—though if the officer is kind enough to say "no," the judge will probably throw him and your case out of his courtroom.

"Would you please explain the purpose of calibrating a radar unit?"

To make sure that it is working properly, of course.

"How often do you normally calibrate the unit?"

The log shows how often he actually did calibrate on the day in question—if his answer agrees with this, you've made your point; if it doesn't, he's digging himself a hole to fall into.

"What method do you use to calibrate the radar?"

Answer should be: he strikes the tuning fork (two forks simultaneously for moving radar) on a non-metallic surface, holds it in front of the antenna, and verifies the speed reading—or both speed readings in the case of moving radar.

"At what time prior to the alleged violation was the radar unit last calibrated?"

You know the answer, you just want to see if he does.

"At what time after the alleged violation was the radar next calibrated?"

Ditto above. Now comes the snapper:

"Officer, isn't it true that proper procedure calls for calibrating the radar with a certified tuning fork before and after an arrest [or whatever the departmental guidelines specify]?"

You will cite the officer's testimony and the calibration log that has been entered into evidence during your motion for dismissal (and if that motion is denied, again during your summation).

What you know and the officer apparently does not, is that both *Wisconsin v. Hanson* and *Minnesota v. Gerdes* establish that proper procedure calls for using the tuning fork within one hour of giving a citation. Two additional cases, *Connecticut v. Tomanelli* and *New York v. Stuck* mention that tuning fork tests were conducted "immediately before and after" the citations.

This same approach is effective if you discover that the radar unit has a record of failures and repairs, and may be presumed to be unreliable; or that the patrol car has had a series of electrical problems, and may not have provided a stable power source for the radar; or that the car's speedometer hadn't been calibrated within the past 6, 12, 18, or however many months prior to the arrest, and so should not have been used with moving radar.

None of these points may be considered crucial enough by the judge to cause him to dismiss your case, but all can help raise "reasonable doubt." Moreover, you may shake the officer's confidence a little, and set him up for when you bring the heavy artillery to bear.

Chapter Nine

THE HEAVY ARTILLERY

Establishing a hypothesis based on typical radar errors.

I t's time to take the officer through the details of the alleged violation. Point by point, you intend to remind him of certain conditions at the time of the arrest; invite him to agree with you on a number of circumstances; ask him to concede that there is room for doubt, however small, that you are guilty of the charge that's been brought against you.

Your line of questioning will be determined by your *hypothesis,* and by the types of radar error that occurred or *could* have occurred in your case.

To that end, let's review the principal sources of error, then see how you'll go about matching them to various theories.

The NBS report on radar failures

Several of the most frequent causes of radar error are listed at the conclusion of the National Bureau of Standards' report on Six Radar Units (see Appendix C). Paraphrased for clarity, they are:

1. Two-piece radars can produce erroneous readings when the antenna is panned across the display console. This is known as "panning error," and would occur if an officer clocking oncoming traffic were to sweep his radar antenna across the inside of the patrol car, intending to clock traffic moving away from him.

2. Air conditioner and heater fans, and alternator or ignition noise can interfere with the radar when no bona fide target is present. This is known as "internal interference," and may cause false readings. If a car should come into the officer's view at approximately the same moment such a reading was obtained, it's conceivable that the radar would lock onto the "speed" of the fan, or whatever, and not the car.

3. Patrol speed "shadowing" can occur during moving radar operations. Whenever there is slower moving traffic ahead of the patrol car, there is the possibility that the radar's low beam component will mistake some other object for the ground. This can result in a lower patrol-speed reading and produce an equivalent increase in the target speed.

4. Target speed "bumping" can occur during moving mode operation. If the patrol car begins slowing down to change directions before the target speed *and* patrol car speed are locked in, some radars may "bump" the target speed by several miles per hour.

5. Transmitting on a police-band or CB radio in the patrol car, or in a nearby vehicle, can cause the radar to produce erroneous readings. This is another source of "interference" which could result in false speed readings being attributed to a target vehicle.

6. The use of the "automatic lock" feature may result in wrong target identification. This is why the NHTSA has recommended disabling auto-lock on the units still in use which have it, and why manufacturers no longer put auto-lock on their equipment.

7. When two or more vehicles are in the radar's beam, it can be difficult to select the correct target. It may be necessary to wait until the vehicles pass by the patrol car, and the one being tracked no longer registers a speed reading, to assure that the correct target vehicle is being tracked. This process, known as obtaining a "tracking history," is one of the most commonly ignored recommendations in training manuals, and probably results in more bad arrests and convictions than any other operator error.

8. External interference from such sources as faulty insulators on high-tension lines, radiation from generators or power stations, radio and TV broadcast transmitters, or military or commercial radar installations may cause faulty speed readings. Because these features are commonplace in the driving environment, most instruction manuals caution radar operators to obtain a "traffic history"—a stable reading several seconds or more in duration to insure the absence of electromagnetic interference—before attempting to measure traffic speeds.

9. Harmonic effects in PLL-type radars may multiply a low (under 20 MPH) speed reading. This may result in a reading twice or even three times as high as the target vehicle's actual speed.

10. The Nichols Effect may occur when moving radar is used in a patrol car traveling through a right-hand bend, reading the speed of an approaching vehicle. In this particular, not uncommon, situation, the angle between the patrol car's *actual* and *apparent* direction of travel will cause the radar to add the value of the cosine error to the target speed.

11. Multi-path radar is a special case, with so many possibilities for error—dual antennas, lane selection, "approach" and "overtake" settings, plus all the regular operating requirements of moving radar—that no officer can be certain that the speed reading he obtained was from the target he intended. (Which will not prevent him saying he is, of course.)

The difficulties involved with getting reliable traffic histories with stationary radar (it is virtually impossible to do with moving radar) also are responsible for many unjustified arrests.

Building a hypothesis

Comparing the mistakes that radar makes with the circumstances in which it is used most often leads to a better understanding of the errors that actually do happen. The most common errors can be grouped into categories according to the mode in which the radar was used and the amount of traffic on the road at the time.

(A)—Moving radar; no other traffic

1. Traffic history error

False speed readings may be produced by faulty power lines or large radio transmitters, or by the officer's own radios or air conditioner. Because the patrol car is moving, it is impossible to test the environment for electromagnetic interference before each speed reading. Unfortunately, such interference occurs intermittently and unpredictably.

> HYPOTHESIS: One of these false signals occurred simultaneously with the appearance of the defendant's vehicle; because there was little or no other traffic in view, the officer naturally mistook the false signal for the defendant's speed.

2. Tracking history error

Police radar is capable of "seeing" good targets—such as large trucks, trains, or even low-flying aircraft—for up to a mile over level, unobstructed terrain. Once the radar has locked onto such a target, it may ignore a passenger car that is much closer to it. Unless the officer "tracks" a car long enough to observe a change in the speed reading as it exits the radar beam, he cannot be certain that the car is responsible for the reading.

Due to the large number of tasks the officer must perform (he must first note a vehicle that is apparently speeding, confirm a speed reading on his radar, verify the patrol car's speed by checking the radar display against the speedometer, complete a tracking history by noting the change in the speed reading and Doppler audio as the target vehicle passes, meanwhile taking care not to alter the patrol car's speed until these tasks have been completed—usually in an elapsed time of under four seconds—while continuing to drive his own car in a safe and responsible manner), it is doubtful that a complete and proper tracking history can be taken with moving radar.

HYPOTHESIS: The officer had insufficient opportunity to perform a tracking history which would have shown that either interference or a better radar target at a distance was the actual cause of the speed reading.

3. Patrol car speedometer error

If the patrol car's speedometer is not calibrated regularly, the prosecution cannot show that the speed reading obtained was free from shadowing error.

HYPOTHESIS: If, according to the maintenance record, the patrol car's speedometer was not calibrated in the six months prior to the arrest, undetected "shadowing error" may have increased the speed reading for the defendant's car.

(B)—Stationary radar; no other traffic

1. Traffic history error

The presence of various sources of electromagnetic interference inside of, or external to, the patrol car creates the possibility of spurious signals.

> HYPOTHESIS: Because the officer did not take a traffic history immediately prior to obtaining a speed reading for the defendant's car, he was not aware of the presence of interference.

2. Tracking history error

If an officer does not complete a tracking history, he cannot be certain that the car closest to the radar is responsible for its speed reading.

> HYPOTHESIS: Because the officer did not track the defendant's vehicle entirely through the radar beam to confirm a change in both the speed reading and the audio Doppler, he mistook a random reading for the speed of the target.

(C)—Stationary "Instant-On" radar; no other traffic

1. Traffic history error

Some police radar units can be operated in a special "instant-on" mode which keeps the unit on standby, emitting no signal, until activated by the officer. Using the radar in this manner effectively precludes the possibility of establishing a traffic history.

HYPOTHESIS: By operating the radar in the instant-on mode, the officer mistook electromagnetic interference—either from external sources or sources inside the patrol car—for the speed reading of the defendant's vehicle.

2. Tracking history error

Instant-on operation increases the possibility that the radar will lock onto the "best" target within its range, not necessarily the one closest to it. Unless the officer performs a full and correct tracking history, he cannot be certain that the speed reading is not for a larger and/or faster target up to a mile away.

HYPOTHESIS: By failing to track the defendant's vehicle all the way through the radar's beam, and confirm a change in the speed reading and Doppler tone, the officer was unable to positively identify the source of the reading.

(D)—Moving radar; other traffic on road

1. Traffic history error

All of the usual sources of interference that affect radar in the moving mode with no traffic also affect it when traffic is present, with the additional problems of shadowing error and greater difficulty in target identification.

HYPOTHESIS: Due to the failure or inability to obtain an adequate traffic history, plus the difficulties of selecting one target from among many, the officer associated a possibly incorrect speed reading with an incorrect target.

2. Tracking history error

Proper tracking procedure is to follow the target vehicle until it passes through the radar's beam and both the speed reading and the Doppler tone change. Depending on how much other traffic is on the road, this task ranges from very difficult to literally impossible.

HYPOTHESIS: The officer was unable to perform a complete tracking history, thus failed to identify the source of the radar reading positively.

3. Shadowing error

According to tests by the National Bureau of Standards, many radar units are extremely prone to shadowing error in the moving mode. Some will produce such error when there is traffic of any type ahead of the patrol car.

HYPOTHESIS: The radar shadowed a slow-moving vehicle ahead of it, under-reported the patrol speed, and caused an increase in the speed reading for the defendant's car.

4. Batching error

Tests by the National Bureau of Standards show that sudden acceleration or deceleration by the patrol car can cause a lag in the radar's patrol-speed reading, and add speed to the target.

HYPOTHESIS: When the officer decelerated to turn around (or accelerated to reach a median crossing in order to turn around) batching error caused the radar to add several miles per hour to the defendant's true speed.

5. Multi-path errors

Multi-path radar units require their operators to select between forward and rearward facing antennas; to choose traffic in the same lane with the patrol car, or in the lane adjacent to it; and to differentiate between traffic approaching the patrol car or in the same lane with it.

HYPOTHESES: The officer confused one of the many possible combinations of settings with the one that would have shown the actual (legal) speed of the defendant's car.

(E)—Stationary radar; other traffic on road

1. Tracking history error

Depending on circumstances, different models of police radar may choose the fastest, the closest, or the largest target when more than one vehicle is present. Unless a full and correct tracking history is performed, the operator cannot be certain that a speed reading was produced by a particular vehicle.

HYPOTHESIS: Due to traffic on the road and the officer's failure to perform a complete tracking history, the wrong speed reading was attributed to the defendant's car.

2. Target direction error

Police radar in the stationary mode cannot tell whether a target is moving toward its antenna or away from it. Thus, when there is both approaching and receding traffic within the radar beam (the most common situation), it's possible to confuse a vehicle moving in one direction with the speed reading of a vehicle moving the other.

HYPOTHESIS: The officer failed to note a faster-moving vehicle in the opposite lane—or on a parallel road beside him—and attributed its reading to the defendant.

(F)—Stationary "Instant-On" radar; other traffic on road

1. Traffic history error

No meaningful traffic history can be obtained when instant-on radar is used. Any form of external and internal interference may be present and go undetected by an operator who switches the unit on only in the presence of a suspected target.

HYPOTHESIS: Internal or external electromagnetic interference was responsible for the defendant's speed reading.

2. Tracking history error

The instant-on feature is no impediment to the performance of a proper tracking history. (Whether it was done or not is another question.)

3. Target direction error

Due to the absence of a continuous traffic history, instant-on radar is more likely to result in target-direction errors.

HYPOTHESIS: The speed reading obtained by the officer was for a vehicle moving in the opposite direction from the defendant, possibly out of the officer's line of vision.

The nitty gritty

Your exact line of questioning and the relative emphasis on individual questions will be influenced by the hypothesis you adopt. Obviously, there's not much point questioning the officer about patrol-speed shadowing and batching if he wasn't using moving radar. Such distinctions aside, it's possible to construct an all-purpose line of questioning that covers most situations.

At this point, it's necessary to touch on a rather unpleasant subject. When you begin asking questions about the circumstances of your arrest, many officers will lie. They may not do so consciously or deliberately—it's really a kind of reflex action. They made the bust, they "know" you're in the wrong; probably they don't remember the details of the arrest, but they'll be damned if they'll admit anything that might be of help to you.

For example, if there was other traffic on the road, don't be surprised if the officer denies it. Stay calm and don't be offended. Did the space on your citation for "Traffic Conditions" list them as "light," "medium," "heavy"? Proceed with:

"Officer, is this a copy of the citation you issued to me on the date in question?" (Show him your copy.)
Yes, it is.

"(To judge) I offer the copy of the citation into evidence. (To officer) Under the space marked 'Traffic Condi-

tions,' the notation is 'moderate to heavy.' Is this your writing?"

Yes.

"Did you so mark the citation?"

Yes.

"Do you still maintain there was no other traffic on the road at the time you took a speed reading for my car?"

When you begin asking questions about the radar's beam width, and so on, don't accept an answer of "I don't know."

"Officer, what is the range of your radar—that is, how far away in feet will it read a target?"

I'm not sure.

"You aren't sure. You did say that you had received training in the use of radar, didn't you? You do have a certificate as a qualified operator, don't you?"

Yes.

"Yet you say you don't know the range of the radar. Then how can you tell, when you get a reading on a vehicle—at any distance whatsoever—that the vehicle is not beyond the range of your radar, and in fact the reading is from something else?"

If he persists in denying knowledge of the range of the radar, you will ask the judge for a "directed verdict of acquittal" based on the incompetence of the prosecution's witness.

The point to all this is, don't just assume that each of your carefully prepared questions is going to receive an honest or forthright answer. It probably won't. You may have to probe, but remember, the truth will come out; the officer can only fob you off to a certain point; after that, it's a matter of perjury, and he knows it. (In fact, after you run him to earth on one question, he may become more co-operative instead of less so.)

Subjects for your cross-examination, in approximate order, are:

1. The radar's range.

2. Its beam width.

3. Tracking history.

4. External interference and traffic history.

5. Internal interference.

6. Panning error.

7. Errors in the moving mode:
   (a) Shadowing error
   (b) Batching error
   (c) Harmonic error
   (d) Moving cosine error
   (e) Multi-path errors

8. Question on the hypothesis.

Having checked the beam width and range of the radar during cross-examination, or in the materials you subpoenaed, you are prepared to home in on the key element of your defense. An example of some questions and answers in a typical defense are:

"Officer, can you tell me the approximate range of your radar unit? That is, how far away will it register the speed of a target?"

The answer should match with the limits of visual observation, about 1,400 feet. If the officer says the range of the radar is *more* than 1,400 feet, then he cannot have followed proper procedure, which is to observe a suspect vehicle apparently traveling at a high rate of speed, and then to confirm the speed with the radar. If the officer says he doesn't know the range of the radar, persist for a while in an effort to get him to commit to some minimum figure.

"From your personal experience in using the radar, would you say that it registers a target up to one quarter of a mile away?"

If the answer is yes, continue with: *"One half mile? One mile? Farther?"* Sooner or later, the officer will put a limit on the radar's purported range, or the prosecutor will interrupt to say that the officer testified that he doesn't know.

"In short, then, your radar could have been reading another target at any distance whatsoever, and you would not have known. Is that correct?"

The answer doesn't matter.

"Approximately how far away was my car when you became aware of a high speed reading?"

Make a note of the answer; it will help you establish whether the officer took an adequate traffic history.

"What did you do then?"

The answer will help you establish whether he performed a proper *tracking* history.

Officer, are you familiar with the term 'beam width?'"

If he isn't, he's in trouble.

"Will you tell me what it is?"

It's the angle between the half-power points of the radar's beam, measured in degrees. In other words, it's the center of the flashlight beam, which "illuminates" the target.

"Can you tell me about how wide the beam of your antenna is at (whatever distance he said he became aware of a high speed reading)?"

Check the answer against Figure 1 on page 13, which shows the beam width (in feet, measured perpendicularly to the centerline of the beam) at various distances from the antenna. If the officer's estimate was in error, or he has admitted not knowing, you'll make an exception to one of life's cardinal rules ("Never wise up a dummy") by revealing the answer to him. This will prepare him, and the court, for the next question.

"At the distance at which the speed reading was first noted, how wide is the radar beam? Approximately how many lanes of traffic does the beam cover at that distance?"

A lane is about 11 feet wide; at any distance beyond 500 feet—one-tenth of a mile—from the radar, the beam will cover all four lanes plus a 50-foot median of a divided highway. The officer's answer, verified in Figure 1, should establish that all traffic in both directions was within the radar beam as the speed measurement was taken.

If the radar was stationary, proceed with:

"In other words, all lanes of traffic in both directions were covered by the beam from your radar at the time you took your speed reading?"

He can only answer yes.

"To your knowledge, does your radar read traffic coming toward you, or going away from you, or does it read it in both directions?"

He should know that it reads both directions.

"Then it would be possible for the radar to give a speed reading for a vehicle traveling the opposite direction from my car, is that correct?"

If he says otherwise, make a note. From here on, assume that you will note *all* factually incorrect answers and refer to them during summation.

"In fact, it's possible that a vehicle ahead of you, going the opposite direction from my car, was the one responsible for the speed reading, isn't it?"

The officer will say that in his opinion your car was responsible for the reading, but don't worry; you're planting seeds of doubt for a later harvest.

If multi-path radar was used in the moving mode, ask:

"Officer, with multi-path radar, what would happen if you used the wrong antenna, or the wrong lane setting, or the wrong directional control? Would it be possible for you to be looking at one car, and for the radar to get a reading from a different car? Are you capable of making such a mistake?"

It's possible, and he's capable.

Tracking the truth

The next set of questions is aimed at establishing how well, or poorly, the officer performed traffic and tracking histories.

"In your training in the use of radar, were you told that certain forms of interference could produce false signals?"

He will say yes, or perhaps, "only in the absence of a valid target."

"For instance, could a high-power utility transmission line interfere with the radar signal?"

Same answer.

"Can airport or military radar interfere with the radar?"
Same answer.

"Have you ever noticed interference from neon signs, street lights, or other sources of high-intensity electrical discharge?"
If he says he hasn't, ask that the radar be plugged in and demonstrate readings from the fluorescent lighting in the room, the tuning fork, or a battery powered watch if you have one. This will demonstrate that radar can produce a speed reading in the total absence of a valid target.
If stationary radar was used, pursue the following line of questioning:

"How long were you at the location where my alleged violation took place before you saw my car?"
If he had just arrived, he cannot have taken a traffic history. If he had been there a while, ask:

"How recently, before seeing my car, had you taken a traffic history?"
If he's confused by the question, explain that a traffic history is the continuous observation of traffic by a police officer, using a radar device, during which the officer has the opportunity to become aware of any form of interference which could affect speed readings.

"I am asking if you performed such a traffic history prior to obtaining my speed reading?"
Whatever the answer, you've begun to establish that radar is subject to false speed readings under certain circumstances—something most members of the public, and quite a few judges, don't know.
If *moving* radar was used, confine your efforts to establishing what a traffic history is; then conclude with:

"Isn't it true that a traffic history, as I have defined it, cannot be taken from a patrol car that is constantly changing its position?"
The answer doesn't matter.

Doppler audio

"Doppler audio" is a tone produced by police radar that rises and falls in pitch with the speed reading. Almost all police radars have Doppler audio (exceptions include older-model CMI Speedguns, and Kustom HR-4 and HR-8), and most manufacturers' manuals advise using it to confirm speed readings. So do at least two court cases: *Akron v. Gray* and *United States v. Fields.* The importance of using Doppler audio is the reason for the following questions.

"Does your radar have Doppler audio—that is, does it produce a continuous audio signal that converts the Doppler shift into an audible tone?"
Probable answer, yes.

"Does the Doppler audio have adjustable volume control?"
Yes.

"Do you normally set the control at full volume?"
Only a hearing-impaired officer—or one addicted to heavy-metal music—would do this.

"At about what volume do you set your Doppler audio?"
Loud enough to hear.

"It's possible to obtain speed readings without the use of the Doppler audio, isn't it?"

It is.

"Do you sometimes operate with the Doppler audio turned off?"
No.

"Can the Doppler audio interfere at times with your use of the radio, or conversation with other officers?"
He's lying if he says it can't.

"Do you sometimes turn the volume down in order to use the radio?"
You may get a yes, but in any case continue with:

"Officer, I realize that it has been several weeks since the alleged violation, and you have written (scores, hundreds, whatever—check his arrest record) of tickets since then; but to the best of your recollection, were you using your radio, and had you turned the Doppler audio down, at any time prior to my arrest?"
You've made it quite clear that he could have; anything less than a flat "no" is a point for your side.

Auto-lock

"Auto-lock" is a feature found on many older radars in use today. It sounds an alarm, and locks in a reading, whenever a pre-set speed reading is reached or exceeded. Many officers still use this feature of their radars, even though the NHTSA, and even radar manufacturers, agree they shouldn't.

"Does your radar have a feature called auto-lock?"
If he is not responsive, describe the feature and what it does, then repeat the question. If the answer is no, skip the next set of questions. If the answer is yes, ask:

"Does your department recommend that you not use auto-lock?"

It should; but whatever the answer, continue with:

"Do you ever use the auto-lock?"
After either answer:

"Are you aware that the use of auto-lock is con-demned by the NHTSA, the International Association of Chiefs of Police, and most radar authorities?"
This may draw an objection as "irrelevant," "incom-petent," or both, but continue with:

"Are you aware that radar manufacturers no longer equip police radar with auto-lock?"
The answer doesn't matter; you've made your point.

Reaching the conclusion

If *moving* radar was used in your arrest, and if condi-tions were such that either shadowing error or batching er-ror *could* have occurred, ask these questions:

"You've testified that, just prior to my arrest, you were traveling (describe direction of travel and route name or number) in (light, moderate, heavy) traffic—is that cor-rect?"
He has so testified.

"Would you describe, to the best of your recollec-tion, the traffic that was on the road ahead of you?"
He may be able to, though chances are he does not recall. Help his memory along with:

"Isn't it true that there was a (car, camper, motor home, trailer or—best of all—a Greyhound bus or 18-wheeler) just ahead of your patrol car?"
If he denies that there was any traffic ahead, remind him that he's already agreed there was other traffic on the

road—suggest that it's been quite a while since the event; that he has written (N number) of tickets since then; and that perhaps your memory of the incident is better than his. If he is adamant, abandon this line of questioning. If he agrees that there was, or could have been, traffic ahead, proceed with:

"Are you familiar with the term 'shadowing error,' and could you please define it?"

If he answers "no," ask:

"Didn't you testify that you took training in the operation of radar?"

Yes.

"And in that training, weren't you told that a vehicle on the road ahead of the patrol car could cause shadowing error?"

He should have been.

"And you were told that, when shadowing did occur, the speed of any target vehicle would be increased in your target-speed window. . .my car's speed, for instance, would read higher than it actually was, is that correct?"

He will have to agree—but will probably add that he was told to verify the patrol-speed reading with the car's speedometer, and that he did this.

If the patrol car's records show that its speedometer hadn't been calibrated for six months or more, this is the perfect spot to insert your questions on that subject—ending with:

"If the speedometer were in error, it would not provide an accurate cross-check with the radar, would it? Some amount of shadowing error could occur, and you would have no way of knowing that, would you?"

If the calibration was not in question, you will continue with:

"Do you check the patrol-speed reading on the radar with the patrol car's speedometer every time you get a high target-speed reading?"

Oh, yes.

The final questioning along these lines is meant to raise the possibility that batching error occurred:

"After you became aware of a high speed reading on your radar—and after you identified the probable target . . . and checked the target speed on the radar . . . and the patrol-car speed on the radar . . . and the speedometer reading, of course—what happened next?"

He slowed down in order to turn around to give pursuit.

"Did you slow down before doing all of the things I just mentioned, or did you do them as you were slowing down to turn around?"

He ought to say he did them all *before* slowing to turn. Continue with:

"Are you familiar with the term 'batching error,' and could you describe it?"

After one or the other of you defines batching error—adding to the target speed by accelerating or decelerating suddenly with the patrol car—continue with:

"So, if you had begun slowing down before doing ALL of the things I just mentioned (run them down again if you think you can get away with it), it would be possible for the radar to add several miles-per-hour to the speed reading for my car?"

Whether he admits it matters not.

As we've seen, checking for shadowing and batching error are only two of the tasks the officer must perform in order to obtain a correct tracking history for moving radar. A slightly different set of tasks is involved in stationary radar—and both types of tracking history are more critical when there is a lot of traffic on the road than when the target vehicle is by itself.

The final line of questioning is designed to show that a full and correct tracking history did not occur (it virtually never does) prior to your arrest, and that as a result there's a high probability that the speed reading the officer obtained was either spurious, or was produced by another vehicle.

"I am going to ask some questions about what I refer to as 'tracking history.' A tracking history is the sequence of events that a radar officer follows in detecting a speed reading and deciding who's responsible for that reading. Do you understand what I mean by this definition of a tracking history?"

He should—though if he wants to call it something else, that's fine.

"Can you describe the tracking history (or whatever) that you made for my arrest?"

If he goes by the book (or the text supplied by the radar manufacturer), he'll say something like: "After visually observing the defendant's car, which appeared to me to be speeding, I verified the speed by means of the radar, etc." Pay special attention to whether he says that he tracked your car "through the radar's beam and observed a change in the speed reading and Doppler audio." (If he actually did this, he's an unusually conscientious officer.)

If he cannot describe a tracking history, or does so improperly, let it stand for now (but make a note).

"When you take a tracking history, you first make a visual assumption about speed, is that right? Do you do this before the radar sounds an alarm—or do you do it after you see a high speed reading?"

This is a toughie; he's on record with the distance he claims to have spotted you speeding, and it's got to be less than the range of the radar. How can he maintain that he made visual identification *before* obtaining a speed reading? (If he was using instant-on radar, forget these questions and concentrate on the absence of a proper *traffic history*.)

If he agrees that he was aware of the speed reading before he identified you as the probable speeder, you've got him:

"In this case, then, you heard the alarm or became aware of the speed reading before you saw my car, is that correct?"

He's already said so.

"Then you did not really take a proper traffic history, did you?"

Bore in with:

"Would the knowledge that the radar had already caught a speeder influence your judgment in making a visual determination of a car's speed? That is, would you be more inclined to agree that a car is going a certain number of miles per hour after the radar had already said it was going that speed?"

He should agree; if not, keep going.

"That's really the purpose of a tracking history, isn't it: to guard against the types of radar error we've talked about by making a visual identification first, then confirming it with the radar?"

Yes.

"But in this instance, you didn't do that?"

No.

At about this point, if all has gone well, or a little later if you still have work to do, you can invite the officer to accept your hypothesis.

"Then it's at least possible that the radar reading you obtained was incorrect or belonged to another vehicle, isn't it?"

Whether he says yes or no, your next line is the one you've been waiting to deliver:

"Thank you, Officer, I have no further questions."

Chapter Ten

THE HOME STRETCH

Completing cross-examination; to testify or not; summing up and motions for acquittal.

T he prosecutor has been seething with impatience to get at the witness and begin repairing the damage you've done. Chances are this is the first time he's run up against a really competent defense to a speeding ticket, and he's not going to be beaten if he can help it. He's going to plug any holes he thinks you may have made in his case, and do his best to refute your hypothesis.

Relax and enjoy life; when he's finished, you get to recross-examine.

While the prosecutor redirects, keep in mind that he is restricted to questions *only* about those points covered in cross-examining testimony, and still is forbidden to ask leading questions. If, for example, you created some doubt about the officer's knowledge of beam width, and the prosecutor now suggests: "During your training, Officer, didn't you learn that the beam width of the Speedgun Six was—" you will be on your feet instantly to object. Because:

"The prosecutor is leading the witness, Your Honor."

Don't be rattled if your opponent succeeds in blunting or refuting one of your key points. Use the time before your recross to think; you may come up with a way of reinforcing your argument.

Suppose you established during cross-examination that there were sources of external interference at the place where the officer took his reading. During redirect, the prosecutor asks how long the officer had been in that position, and if any indications of interference were observed during that time (in effect, was a proper traffic history performed?) The officer's answer appears to rule out any possibility of external interference. When it's your turn again, you'll ask:

"Do you have a schedule or timetable for when the (power station, radar installation, or whatever) is turned off or on? That is, do you have any way of knowing exactly when it goes into operation?"
It's most unlikely he will know this.

"Then it's possible the (whatever) wasn't in operation when you took your traffic history, but was turned on between that time and the time you obtained my speed reading, isn't it?"
He should answer that it's possible.
Again, remember that the prosecution is barred from introducing any subject that was not covered in either direct or redirect examination. Should he attempt to do so, either at this time or during his summation, you can ask for a directed verdict of acquittal, or use the transcript of the trial to lodge an appeal.

Motion to dismiss

You'll know the floor is yours when the judge asks something like, "Does the defense wish to testify or call witnesses?" Whether you wish to do either, first you intend to address some motions to the court.

Actually, of the several types of motion for dismissal or acquittal, one can be made (in writing) even before you come to court. If, as a result of your pre-trial research, you have found that the area where you received your ticket was an illegal speed zone, as defined by the traffic regulations in your state—or if the road wasn't properly posted, or the date written on your ticket was incorrect, or some similar error—you can ask for a dismissal, and should get one.

If, by your trial date, the police still haven't come through with the items named in your subpoenas, you can move for an acquittal or a continuance on the basis of "non-disclosure."

After the main part of the trial has taken place, and the prosecution rests, you will move for a "directed verdict of acquittal" based on one of two premises: (a) the failure of the prosecution to make its case, or (b) procedural errors that have prevented or prejudiced your ability to present a defense.

If you've spotted an error in the presentation of the prosecution's case, begin by saying something like this: "Your Honor, the law (recite the appropriate statute from your research) requires that signs be posted on any highway where radar is used. The prosecution failed to establish that signs were posted in the area where I received my citation (it doesn't matter if signs were there or not if the prosecutor failed to have the officer *say* they were). The requirements of the law not having been met, I move for acquittal."

The same basic motion is in order if the prosecutor failed to mention the legal speed limit, or if the date of the offense or other necessary information was missing or incorrect on the citation.

If any of the items you subpoenaed were not produced, offer a motion to dismiss based on the fact that you were "denied access to evidence necessary to the defense, which might have shown that the radar reading was incorrect or improperly obtained." Repeat the motion for each of the subpoenaed items that was not made available to you.

If it appears that the officer's training was inadequate, or if he goofed in citing the range or beam width of his radar, you can move in with: "Your Honor, during questioning, the officer admitted not knowing the range or beam width of the radar (or incorrectly estimated the range and beam width). Because accurate knowledge of range and beam width are essential to target identification, it is apparent that the prosecution's witness is not qualified. I move that the charges be dismissed on the basis that the witness is incompetent."

The officer's testimony may have revealed some clear-cut mistake in the way he used the radar, or the circumstances surrounding its use. Sometimes, this kind of information is best saved for your summation, where the object is to stress that "reasonable doubt" exists.

If the error is contrary to documented proofs, it may support a motion for acquittal. Suppose the officer has testified that he calibrated the radar at the beginning and end of his shift, and that according to departmental policy this is sufficient to insure the accuracy of the unit. You can offer a motion to acquit on the grounds that the officer's procedures were inadequate according to *Minnesota v. Gerdes, Connecticut v. Tomanelli* and *New York v. Stuck* (see Appendix A).

The defendant for the defense

Once you've exhausted your motions to dismiss, and provided the judge has denied them, you will have a chance to call witnesses and take the stand yourself. Or you may bypass these options and proceed directly to your summation. If you decide on the latter course, say, "The defense rests, Your Honor, but I would like to reserve the right to make a closing statement at the appropriate time." Now is also the time to renew your motion to acquit.

If you decide to call a witness—typically someone who was in the car at the time of the arrest—you will identify the witness and he or she will be sworn in. Begin by asking the question, *"Were you present at the time and date in question?"* and go on with questions designed to establish the fact that you were not speeding. Remember that in "direct" you cannot lead the witness; you must ask only questions that do not suggest a response (just the facts, ma'am). A useful approach might be:

"Where were you on the date in question?"
Riding along with you in your car.

"Was there anyone in the car with you besides the defendant?"
Whatever.

"Where were you seated in the vehicle?"
The right front seat, let's say.

"What happened at approximately (the time of the arrest)?"
I became aware that we were being followed by a police car.

"How did you become aware of that fact?"
I heard the siren, and you—uh, the defendant remarked that there was a police car behind us.

"Did anything occur that called your attention to the speed of the defendant's car?"
Yes, he asked me to look at the speedometer.

"Did you look?"
Yes.

"Was it registering a speed reading?"
Yes.

"What was that speed reading?"
Fifty-five miles per hour.

"Had the defendant slowed down before asking you to look at the reading?"
No.

"What happened next?"
Presumably, the witness will say that you did not change speed just prior to being stopped, that your speed was within the posted limits, and that the speed the officer cited you for was incorrect. He may be able to add that the officer did (or did not) invite you to look at the radar reading; refused to answer your questions; agreed your speed might not have been the one on his radar, but that you were going too fast anyway—anything that might cast doubt on the validity of the arrest, so long as his testimony is the truth.

When you've completed your direct, the prosecutor can cross-examine. Now it's his turn to ask leading questions, and he may use them to impugn the witness's honesty (by suggesting he's lying to help out a buddy); grill him on his ability to estimate speeds; rake him over the coals about his version of events, and generally make life unpleasant.

If he does all this, don't waste time making pointless objections. Unless he pulls out a rubber hose and starts using it, the judge probably won't intervene, so save your energies for redirect. That's when you'll quietly and calmly go over the key elements of the witness's testimony to reestablish the main points, just before announcing: "No further questions."

If you do decide to testify yourself, the ground rules and likely sequence of events will be approximately the same as for your witness. You'll be sworn in, take the stand, and inasmuch as there's no one to ask you questions, you'll give your version of what happened in simple narrative. You may read it, if the rules of the court allow (better than freezing up and forgetting what you were going to say); but you'll score better with a jury if you deliver it from memory or a few notes.

The prosecutor will cross-examine, doing his best to confuse you and lead you into conflicting statements or apparent contradictions. Remember, it's not the truth he's after, it's your hide. Take your time, think about each question before answering, and stick to the bare facts. Don't rise to suggestions about what *might* have happened; all you know is what did happen. You were driving at a legal speed when you were stopped and cited for going faster; you assume that a false speed reading or incorrect target identification was to blame. Period.

Usually, if there's no jury, you're better off not testifying. What's the point? You have little to say other than that you weren't speeding, so why subject yourself to cross-examination by someone who's trying to make you appear guilty? Moreover, if you're defending yourself against a speed trap where you were exceeding an unreasonable limit, you can avoid perjuring yourself or pleading guilty by deciding to testify.

Remember, legally you cannot be required to testify, and the judge *must not* infer anything from your decision. If he suggests that not testifying implies guilt (unlikely, but not impossible), ask that his statement be made a part of the trial record—creating a strong argument for reversal if you appeal.

Summing up

The time has come for your closing statement. Time to pull together the important pieces of the case; to arrange them in logical order; to build a foundation of proof upon which to erect a monument to your innocence, a shining edifice of truth and justice and—

Well, don't get carried away.

Facts are more important than eloquence, and we're assuming the facts are on your side. The objective during summation is to cite testimony that supports your hypothesis and casts doubt on the prosecution's version.

Here again, as with your initial preparations, it would be nice to have help from an attorney, someone to analyze the testimony and search the law for precedents. But if you've come this far alone, you're better equipped than before to do a little research and apply the results during summation.

The first step is to review and evaluate what has been learned from the physical evidence and testimony. Do the facts support your version, or has information emerged which requires you to modify or abandon your original hypothesis? Don't be afraid to change your mind now that you're in possession of all the evidence; during summation, you are allowed to call attention to whatever facts you wish—suggest how they should be interpreted; point out flaws or gaps in the prosecution's case; offer references showing that proper procedures weren't followed, or the officer's training was inadequate, or the radar was suspect —whatever is necessary to undermine the prosecution's case and establish "reasonable doubt."

Format for summation

The following format summation can be adapted to suit the circumstances of your particular case. This example was made before a judge; if the argument is to be made before a jury, citing case law usually is ineffective:

Your Honor, the evidence has shown that the speed reading obtained by the radar was either incorrect, or was taken from another vehicle.

The officer has testified that he did not take a full and proper tracking history, as recommended by the federal (NHTSA) standards and the operator's manual for this radar. Since the officer has admitted that he did not track my car through the radar beam and confirm a change in the audible tone, a complete tracking history was not performed.

The testimony also shows that there was a military radar (airport radar, power lines, a power station, flashing neon signs, a radio or TV transmitter) in the area that could have affected the radar's reading. The officer has testified that he did not take a traffic history immediately before getting a speed reading (could not have taken a traffic history due to the fact that he was using moving radar), and so could not have been completely aware of the effects of external interference.

The officer has testified that he was talking on his police-band radio when he took the speed reading. Tests (by the National Bureau of Standards) have shown that police-band radio can and does produce speed readings on traffic radar.

The officer has testified that his formal training in radar was less that 24 hours classroom and 16 hours supervised field experience. This is less than the *minimum* training considered adequate by authorities (NHTSA or IACP). Because of the many possible sources of error, radar readings are not reliable as evidence unless taken by an operator with at least as much training as called for by the IACP.

The testimony shows that there was traffic on the road ahead of the patrol car which could have caused shadowing error. The testimony also shows that the officer began braking before he completed his tracking history and locked in the target speed. In both of these cases, speed can be added to the target vehicle. The officer has testified that several of those things were, or might have been, present when he took his speed reading.

The K-55 radar used in this case was found to have the widest beam width of any device tested by the National Bureau of Standards. Excessive beam width is known to cause problems with target identification, as can excessive range."

Add a reference to any errors or insufficiencies that you discovered in the subpoenaed materials at this point. Then:

Your Honor, for all of these reasons I submit that the prosecution has not proved beyond a reasonable doubt that I was speeding, and I respectfully ask the Court for a verdict of 'not guilty.'

ONE FOR THE ROAD

You can sit down, but don't get comfortable just yet. The prosecutor gets the final word. Usually, he will simply go over the elements of the citation, asserting that each has been proven, and ask for a guilty verdict. Sometimes, if he is well enough versed on the subject, he may throw *Honeycutt* at the judge; this is not a flower or part of a beehive, it's a legal case.

Honeycutt v. Commonwealth of Kentucky is very popular with prosecutors, enforcement agencies and radar manufacturers, because it says, in part, "a properly constructed and operated radar device is capable of accurately measuring the speed of a motor vehicle," and adds: "The courts will treat as sufficient evidence of a radar's accuracy uncontested testimony that the unit was tested within hours of its specific use and found to be accurate..."

Honeycutt goes on to say that "a few hours" instruction normally should be enough to qualify an operator, and, perhaps most tellingly, that the radar's target vehicle is "the one out in front, by itself, nearest to the radar unit."

This little list of horrors was perpetrated back in 1966, which is perhaps its only excuse, considering that every point (except the first, that radar is capable of measuring speed) is factually and scientifically wrong!

By now, the number and variety of mistakes that traffic radar can make, and the consequent need for adequate operator training, are well understood. No one any longer claims that a tuning fork test—particularly one made within a "few hours" of use—can guarantee the accuracy of the unit. And the part about "out front, alone, nearest the radar" has probably misled more police officers, and resulted in more bad arrests, than any other single court decision.

Although the courts should recognize that *Honeycutt* is rather an old decision (it was handed down 15 years before either the NBS tests or the IACP results were available), it continues to be cited in support of questionable radar evidence.

Fortunately, the language in *Honeycutt* can be interpreted to leave room for the kind of defense you've developed while reading this book. If during his summation the prosecutor refers to *Honeycutt* and claims it rules out the arguments you've made—specifically, your allegations of insufficient operator training and incorrect target identification—you will have, or can ask for, an opportunity to add to your summation. Depending on the part or parts of the decision cited by the prosecutor, you can answer:

"Your Honor, although the decision in *Honeycutt* says that a 'properly constructed and operated radar device is capable of accurately measuring speed of a motor vehicle,' it goes on to say that the courts *will not* take judicial notice of the accuracy of *a particular radar device*. The testimony has shown that the 'particular device' used in this case did not produce an accurate speed reading." Or. . .

"While *Honeycutt* says that 'a few hours' instruction *normally* should be enough to qualify an operator, some circumstances of the present case were not normal (internal or external interference, other traffic, absence of traffic or tracking history, shadowing, bumping, or anything else you can think of). Since *Honeycutt*, both the National Highway Traffic Safety Administration and the International Association of Chiefs of Police have declared that 24 hours classroom and 16 hours supervised field experience is the *minimum* standard for adequate training." Or. . .

"Although *Honeycutt* stated that the target vehicle is 'the one out in front, by itself, nearest to the radar,' it also stipulated that the radar device itself be 'properly constructed and operated.' Some years after *Honeycutt*, the National Bureau of Standards tests showed that many radar units did not meet performance standards recommended by the federal government and adopted by the Florida Radar Commission, among others.

"Moreover, the testimony in this case shows that the radar was not 'properly operated'—that is, a correct tracking and traffic history were not obtained by the officer. Under these circumstances, it is well known that radar may register the closet, the fastest, OR the largest target within its range.

"I respectfully ask for a verdict of 'not guilty.'"

The verdict

Was the verdict guilty or not? Did you win or lose? Was it the lady or the tiger? Actually, the protagonist of the classic short story had a third choice, which was to open neither door, and presumably starve. You can choose not to fight the ticket you didn't deserve, or that never should have been issued; but if you do, you reward injustice, promote further abuses, and help perpetuate a system that trades safety for financial gain.

Better the tiger.

If you got a good judge, one who was really interested in your case and willing to decide it on the merits— or if the officer proved cooperative and honest (or maybe didn't show up for the trial)—or if the arrest was weak and you did such a *damned* good job they had to let you off, then the lady (for "Justice" is one) is yours.

If you lost, but in the process had your say, and tried your best, and ended up feeling you got a fair shake—maybe it was worth the time and trouble anyway.

At least they didn't walk over you; you made them earn their money. If enough people did the same, the profit motive would go right out of indiscriminate ticketing and frivolous enforcement, and we might see radar used as it should be—to back up arrests for gross violations, and as a means of obtaining speed readings without high-speed chases.

Should you appeal?

If you really got ripped off, and are in no mood to be philosophical about it, you may want to consider an appeal. Traffic convictions aren't often appealed because of the time and costs involved; but the major portion of radar case law has been created in appellate courts. So if you believe the prosecution's evidence was insufficient, or that you were denied the right to present a defense, or that your trial was unfair or prejudiced for any reason, you may be entitled to another one.

The type of appellate procedure in your state may affect how you handle your defense. In a few states, the lower courts are not courts of record. If you lose in such a court, you may have an automatic right to a new trial (trial de novo) at the next level. This is ideal, since you can treat the first trial as a practice round.

The most common form of appeal is a review of the trial record to determine if the court committed any errors that prejudiced the defendant. Remember that only issues raised on the record during the trial can be appealed. For example, if you move to suppress evidence derived from a particular radar device, and the judge denies your motion, you can appeal that ruling. If the judge was never *asked* for a determination, you could not appeal.

To appeal, you must have a written transcript or tape recording of your first trial. Most courts record their proceedings as a matter of course, but if you don't see a recorder on duty when your trial begins, you should ask for one. Generally, a judge will not refuse this request; if it happens, you can base an appeal on variations in trial procedure which the judge kept from being recorded.

A transcript will cost you several hundred dollars, a copy of the tape only $20–$40, so start with the tape to see where you stand.

Even if you handled your first case without a lawyer, you probably should hire one for an appeal. This is because appeals are decided more on points of the law involving procedural error than on the original question of guilt or innocence. You may be at a considerable disadvantage arguing an appellate action *pro se*.

A great number of documents, copies and records must be prepared and filed by the deadline for your appeal, which usually is about 30 days from the date of conviction. These include all the documents involved in the case, such as copies of the citation and the subpoenas, copies of all the exhibits, and a transcript of the testimony. The defendant also is required to file a "memorandum of law" spelling out his claim that an error of law has been committed.

If you really want to do-it-yourself on appeal, return to your (by now) good friend, the Clerk of the Court, and ask for help in filing the documents and following proper procedures.

After that, you wait. The appellate court will hear the case if it feels argument is necessary; otherwise, it may simply review your memorandum, and the one filed by the prosecutor, before deciding to uphold or reverse the conviction. Either way makes a good story the next time you're with friends.

A different kind of summation

Whether or not you win your appeal—or if you already won at your first trial—or even if you haven't been to court and are just trying the idea on for size—it's nice to know that as a result of reading this book you qualify as one of the best-informed motorists in the country on radar enforcement and the law.

Your new knowledge will help protect you from radar speed traps, and better equip you to defend yourself if you are caught in one. In the long run, such knowledge may help cure enforcement and judicial authorities of the delusion that radar is always right.

Today, well over 90 percent of radar cases result in convictions, in spite of warnings by the experts that radar frequently is wrong. It would be a pleasant change, and a benefit to all concerned, if the innocent had at least a fighting chance at acquittal. TV newsman, Harry Reasoner, tells a story about a friend to whom he once remarked philosophically, "Well, Joe, you win some and you lose some."

The friend though a minute, then said, "Gee, that would be nice."

Indeed it would.

SIGNIFICANT RADAR CASE LAW

The following excerpts from case law are examples of how various courts have viewed radar evidence. It is not recommended that the non-professional attempt to cite cases and precedents as part of a defense against a radar speeding ticket; but for those who have the time and interest, research of the law in one's own state will provide an overview and valuable insights into how such cases have been handled in the past. Such research may even suggest which lines of defense are most likely to succeed in a particular jurisdiction.

Allweiss, State v. (1980 FL)

The Pinellas County Court found that testing procedures for radar equipment are legally insufficient. "The use of such a tuning fork furnished by the manufacturer in this court's opinion is tantamount to allowing the machine to test itself. A tuning fork furnished by the manufacturer is but an extension and part of the total speed measuring apparatus which is furnished by the manufacturer upon delivery."
Not guilty.

Aquilera, State v. 711-101S Dade County (1979 FL)

The Dade County Court sustained a Motion to Suppress the results of radar units in the prosecution of 80 speeding violations. After more than 2,000 pages of testimony and arguments and 33 exhibits, the court decided that the reliability "of the radar speed measuring devices as used in their present modes and particularly in some cases, has not been established beyond and to the exclusion of every reasonable doubt, nor has it met the test of reasonable scientific certainty."

Not guilty.

Dantonio, State v. 115 A 2d 35 (1955 NJ)

The Supreme Court of New Jersey held that evidence based on radar readouts could be sufficient if it is shown that the radar was properly set up and tested by police officers. There is no need for expert testimony.

Radar speed readings are not conclusive, but merely admissible evidence to be weighed with other evidence; the possibility of errors in the operation of the radar does not affect admissibility but simply affects the weight of this evidence.

Judgment affirmed.

Edwards, State v. 80 03 0390 (1980 DE)

The Court of Common Pleas found evidence based solely on a reading from a K-55 moving radar not sufficient for conviction because this unit has not proven trustworthy or reliable.

Not guilty.

Fields, United States v. 3-81-0064M (1982 OH)

The District Court held that it was impossible for the court to determine from the radar results whether the defendant was traveling 43 MPH or whether the Speedgun Eight was measuring the rotation of the ventilation fan at the sewage pumping station next to the officer's car.

It also was determined that the officer was not qualified to operate the device because he was not aware of the conditions for correct operation. He neither calibrated nor operated the radar unit properly.

The court found that "the expert testimony presented by the prosecution in this case raises the issue as to whether any type of radar device is scientifically reliable for the purpose of detecting the velocity of an automobile traveling on a highway where there are reflecting objects, electromagnetic interference or other vehicles within the beam range of the device."

Not guilty.

Gerdes, State v. 191 N.W. 2d 428 (1971 MN)

The Supreme Court of Minnesota ruled that courts may take judicial notice of the underlying principles and reliability of properly tested and operated radar devices without requiring expert testimony concerning the theory and mechanics of a particular unit. However, where the only means of testing the accuracy of a radar device is an internal mechanism which is an integral part of the unit, and there is no evidence other than the radar reading that a motorist was driving at a speed in excess of the limit, the conviction cannot be sustained.

The requisite conditions for proving the accuracy of a particular instrument are:

1. The officer reading the device must have adequate training and experience in its operation;

2. The officer should testify about how the unit was set up and the conditions under which it was used;

3. A showing must be made that the machine was operated with a minimum possibility of distortion from such external interference as noise, neon lights, high-tension power lines, high-power radio stations, and other similar influences;

4. On the occasion when the machine is set up, its accuracy must be tested in some external manner by a reliably calibrated tuning fork or by an actual test run, using another vehicle with an accurately calibrated speedometer.

Reversed.

Hanson, State v. 270 N.W. 2d 212 (1978 WI)

The Wisconsin Supreme Court revoked the *prima facie* status of radar and set minimum conditions for the introduction of radar evidence. Judicial notice may not be taken of the accuracy and reliability of moving radar in view of credible, conflicting expert testimony concerning the accuracy of the device.

Courts may take judicial notice of the reliability of the underlying principles of speed radar detection that employ the Doppler Effect as a means of determining the speed of moving objects.

Prima facie presumption of accuracy sufficient to support a speeding conviction will be accorded to moving radar upon testimony by a competent operating police officer that:

1. He had adequate training and experience in its operation;

2. The radar device was in proper working condition at the time of the arrest;

3. The device was used in an area where road conditions were such that there was minimum possibility of distortion;

4. The input speed of the patrol car was verified, the car's speedometer was expertly tested within a reasonable period following the arrest; and

5. Such testing was done by a means which did not rely on the radar device's own internal calibration.

Reversed and remanded.

Honeycutt, Commonwealth v. 408 S.W. 2d 421 (1966 KY)

The Court of Appeals of Kentucky found that courts will take judicial notice of the fact that a "properly constructed and operated radar device is capable of accurately measuring speed of a motor vehicle." But the courts will not take judicial notice of the accuracy of a particular radar device.

The courts will treat as sufficient evidence of a radar's accuracy uncontested testimony that the unit was tested within hours of its specific use and found to be accurate by use of a calibrated tuning fork and by comparison with the speedometer of another vehicle driven through the radar field.

Operators are not required to understand the scientific principles of radar but need to be able to properly set up, test, and read the instrument; a few hours' instruction normally should be enough to qualify an operator.

The radar's target vehicle is the one out in front, by itself, nearest to the radar unit.

Judgment affirmed.

Oberhaus, State v. TR 82-4099 (1983 OH)

The Municipal Court sustained a Motion to Suppress the results of the K-55 radar in the moving mode. The court granted judicial notice to the K-55 in the stationary mode only.

Not guilty.

Perlman, People v. 392 N.Y.S. 2d 985 (1977 NY)

The Suffolk County District Court ruled that the digital radar device used to apprehend the defendant was not proved to be accurate or adequately tested because no external test had been performed before or after the arrest; there was no record of testing, no proof of periodic testing by expert technicians, and no scientific testimony about the reliability or fallibility of internal calibration.

Defendant adjudicated not guilty.

Tomanelli, State v. 216 A. 2d 625 (1966 CT)

The Supreme Court of Connecticut found that the Doppler Principle is a proper subject for judicial notice. This judicial notice does not extend to the accuracy or efficiency of any given police radar instrument.

The court also noted that "outside influences may affect the accuracy of the recording by a police radar set sufficient to raise a doubt as to the reliability of the speed recorded."

The tuning forks used to test the radar unit must be shown to be accurate to be accepted as a valid test of the unit's accuracy. To establish the accuracy of a radar unit the operator must testify:

1. That he made tuning fork tests before and after the motorist's speed was recorded;

2. That these tests were made by activating what were described as 40-, 60-, and 80-MPH tuning forks, and by observing in each test that the unit indicated corresponding readings; and

3. That no effort is made by the defendant to attack the accuracy of the tuning forks.

No error.

INVENTORY OF POLICE TRAFFIC RADAR

The following radar units, all included in the International Association of Chiefs of Police (IACP) tests and contained in that agency's "Consumer Products List," account for at least 90 percent of the models currently in use on American highways.

The fact that certain models have not changed names over the years, while their circuitry and performance characteristics have, complicates the task of demonstrating possible errors by the radar. If the unit used against you was manufactured prior to January 1984 (the date of the IACP report), you should subpoena any document the department has pertaining to beam width, range and other performance data for the radar.

The comments on each of the models that follows, unless attributed to the National Bureau of Standards (NBS) or International Association of Chiefs of Police (IACP) tests, are the authors'.

Asterisks (*) identify radar models no longer in production, but presumed still in use. Beam width and range are given where known. For test results on several of the units, see Appendix D "Extract From National Bureau of Standards."

Broderick Enforcement Electronics (B.E.E.)
7155 Antigua Pl.
Sarasota, FL 33581

Models:
BEE-36, K-band, moving.

> Beam width: 15 degrees.
>
> In initial IACP tests, the tuning fork calibration certificate was missing; unit was sensitive to high and low temperatures and humidity; speed display readability and internal circuit not up to specifications; signal processing channel sensitivity not up to specifications in stationary mode; low voltage supply, signal processing channel sensitivity, and target channel speed display did not comply with manufacturer's specifications.

BEE-36, X-band, moving.

> Beam width: 18 degrees.
>
> Initially, the IACP reported tuning fork calibration certificate missing; sensitive to low voltage supply; signal processing channel sensitivity in stationary modes and target channel speed display in moving mode not up to specification; target channel speed display not in compliance with manufacturer's specifications.

CMI, Incorporated
P.O. Box 38586
Denver, CO 80238

Models:
Speedgun One, X-band, stationary*
JF-1000*

> Beam width: 16 degrees
> Range: 1,500 ft.

The oldest of the Speedgun series; manufacturer makes no mention of range in any of its literature; lacks audio Doppler; lacks range control.

Speedgun Three, X-band, stationary*
Speedgun Five, X-band, moving & stationary*
Speedgun Six, X-band, moving & stationary*

Beam width: 16 degrees

Range: 1,500 ft.

The manufacturer publishes no information about the range of these units, so the officer lacks proper reference for interpreting speed readings; no audio Doppler; unit has auto-lock.

Speedgun Six*

Beam width: 18.8 degrees

According to the NBS report, this unit is subject to severe shadowing errors, power surge, and interference from CB and police-band radios.

Speedgun Eight, X-band, moving & stationary*

Beam width: 18.6 degrees

Designed primarily to be used in the moving mode: has no audio Doppler; no published information on range; has auto-lock. According to NBS was subject to shadowing error and power surge.

Speedgun Magnum, X-band, moving & stationary

Beam width: 18 degrees

In IACP tests, sensitive to low temperature and low voltage supply; contrast of speed display not up to specifications; signal processing channel sensitivity in stationary mode not up to specifications.

DECATUR ELECTRONICS, INC.
715 Bright St.
Decatur, IL 62522

Models:
Rangemaster 715, X-band, stationary*
Rangemaster 715, X-band, moving*

> Beam width: 16–24 degrees
>
> Range: 2,500 ft.
>
> Claimed range: 5,000–7,500 ft.
>
> Claimed range far exceeds human capability for target identification; among widest of all radar beam widths; has auto-lock.

MVR-715, X-band, moving

> Beam width: 17.5 degrees
>
> In NBS tests was subject to shadowing error, panning error, power surge and internal interference from heater fan, ignition and CB radio. IACP tests showed it sensitive to low temperatures and subject to internal police-band interference.

MVR-724, K-band, moving

> Beam width: 15 degrees
>
> In IACP tests, contrast of speed display not up to specifications; subject to internal interference from CB.

RA-GUN KN-1, K-band, stationary

> Beam width: 15 degrees
>
> In IACP tests, sensitive to low voltage supply, interior police-band radio and exterior CB interference.

RA-GUN GN-1, X-band, stationary

Beam width did not comply with specifications in IACP tests; subject to malfunctions caused by high and low temperatures, humidity and low voltage supply; signal processing channel sensitivity not up to specifications.

KUSTOM ELECTRONICS, INC.
8320 Nieman Road
Lenexa, KS 66214

Models:
MR-7, X-band, moving*
MR-9, X-band, moving*
TR-6, X-band, stationary*

Beam width: 12 degrees

Range: 1,800 ft.

Some 15,000 of these units are in use. Operator manuals admit to spurious readings, as does the company's letter listing sources of interference. No patrol speed display; has auto-lock and audio alarm.

MR-7, X-band*

Beam width: 14.3 degrees

In NBS tests, subject to panning errors, power surge and internal interference from police-band radio.

MR-9, X-band*

Beam width: 13.3 degrees

In NBS tests, subject to panning error, power surge, internal and external interference from CB radio.

KR-10, K-band, moving
KR-11, K-band, moving

Beam width: 12 degrees

Range: 1,800 ft.

Claimed range: 4,000 ft.

KR-10 is the stripped-down version of the K-11. Claimed range greatly exceeds the visual tracking ability of any operator; auto-lock increases chance of target identification error.

KR-10-SP, K-band

Beam width: 15 degrees

In IACP tests, signal processing channel sensitivity not up to specifications; antenna near-field power density exceeded manufacturer's specification; signal processing channel sensitivity and target channel speed display did not comply with manufacturer's specifications.

KR-11, K-band

Beam width: 15 degrees

Range: 4,100 ft.

In IACP tests, sensitive to low temperature; Doppler audio and contrast of speed display not up to specifications; signal processing channel sensitivity in both stationary and moving modes not up to specifications; target channel speed display in stationary mode, high speeds, not up to specifications; patrol channel display for tracking not up to specifications.

HR-4, K-band, stationary*
HR-8, K-band, stationary*
HR-12, K-band, moving*

Beam width: 12 degrees

Range: 2,000 ft.

Circuitry similar to KR-11. Has auto-lock, audio alarm, phase-lock loop (PLL), and Doppler audio.

HR-8, K-band, stationary

In IACP tests, beam width did not comply with specifications; sensitive to low temperatures, low voltage supply and external police-band radio interference; antenna near-field power density exceeded manufacturer's specifications.

HR-12, K-band, moving

Beam width: 15 degrees
Range: 2,640 ft.
In IACP tests, sensitive to high temperatures and humidity; signal processing channel sensitivity in stationary mode not up to specifications; antenna near-field power density exceeded manufacturer's specifications.

Falcon, K-band, stationary

Beam width: 15 degrees
Range: 2,500 ft.
In IACP tests, signal processing channel sensitivity and target channel speed display did not comply with manufacturer's specifications.

Road Runner, K-band, stationary

Beam width: 15 degrees
In IACP tests, did not comply with signal processing channel sensitivity; test could not be completed because of technical problems in the Road Runner.

Trooper, K-band, moving

Beam width: 15 degrees

In IACP tests, contrast in speed display not up to specifications; antenna near-field power density exceeded manufacturer's specifications.

H.A.W.K., X-band, moving

Beam width: 12 degrees

Range: 1,500 ft.

An example of "multi-path" radar with forward- and rear-facing antennae, direction and lane settings. Operator must have all settings correct or target errors will occur.

M.P.H. INDUSTRIES, INC.
15 S. Highland
Chanute, KS 66720

Models:
K-15, K-band, stationary

Beam width: 15 degrees

In IACP tests, tuning fork calibration certificate missing; subject to malfunctioning due to vibration; speed display lacked specified contrast; signal processing channel sensitivity not up to specifications; low voltage supply did not comply with manufacturer's specifications.

K-15, X-band, stationary

Beam width: 18 degrees

In IACP tests, tuning fork calibration certificate missing; sensitive to high temperatures; Doppler audio not up to specifications; poor contrast on radar readout; internal circuit not up to specifications; low voltage supply did not comply with manufacturer's specifications.

K-35, K-band, stationary

Beam width: 18 degrees

In IACP tests, tuning fork calibration certificate missing; subject to malfunctioning due to low and high temperatures and humidity; test could not be completed because of technical problems in the K-35.

K-35, X-band, stationary

Beam width: 18 degrees

Tuning fork calibration certificate missing; audio Doppler and audio alarm not up to specifications; speed display readability and signal processing channel sensitivity not up to specifications; subject to external interference from CB radios.

K-55, K-band, moving

Beam width: 15 degrees

In IACP tests, tuning fork calibration certificate missing; speed display readability and signal processing channel sensitivity in stationary mode not up to specifications; low voltage supply did not comply with manufacturer's specifications.

K-55, X-band, moving

Beam width: 20.4–24.6 degrees

Range: 5,200 ft.

Cheapest unit on the market in 1979, sold for as little as $375 (in New Jersey, in 1978); impossible to tell early units from later, supposedly improved ones; plastic-cased components subject to overheating; rejected by Washington State Police after tests at University of Washington; has auto-lock, audio alarm, no range control. In NBS tests, two K-55s proved subject to shadowing error, panning error, power surge, internal and external interference from CB. In IACP tests, tuning fork calibration certificate missing; beam width greater than specifications; readability of speed display not up to specifications; signal processing channel sensitivity and target channel speed display not up to specifications; sensitive to vehicle's alternator; test of other interior and external interference not completed due to unusually low power density.

S-80, K-band, moving

Beam width: 15 degrees

In IACP tests, tuning fork calibration certificate missing; beam width greater than specifications; readability of speed display and internal circuit not up to specifications; target speed display did not comply with manufacturer's specifications.

S-80MC, K-band, moving

Beam width: 15 degrees

In IACP tests, tuning fork calibration certificate missing; sensitive to low temperatures, humidity and vibration; tests terminated due to failure of right digit in target display.

S-80MC, X-band, moving

Beam width: 18 degrees

In IACP tests, tuning fork calibration certificate missing; sensitive to humidity; readability of speed display and internal circuit not up to specifications; subject to interior police-band and CB interference; target channel speed display did not comply with manufacturer's specifications.

Authors' Note

In interpreting the above data and comments, it should be borne in mind that the IACP tests were to all intents and purposes a fraud on the public. The agency's clear intention was to stifle growing criticism of the inadequacies of police radar in general, and some of its shoddier examples in particular, by conducting product tests that were essentially meaningless, and compiling an "approved product list" from which no radar unit, however inadequate, could be excluded.

In order to achieve these goals, manufacturers were allowed to supply hand-picked units for testing; where these proved inadequate, they were modified by the manufacturers, then re-tested to insure approval.

In spite of such precautions, some units still were unable to complete the test procedures satisfactorily—at which point the IACP simply declared them in compliance and added them to the approved list anyway.

Subsequent to its radar test program, the IACP came under investigation for alleged tax fraud and several other federal violations unbecoming to officers of the law.

TESTING OF POLICE TRAFFIC RADAR DEVICES: EXTRACT OF IACP REPORT

Test Program Summary, April 1984
International Association of Chiefs of Police

Executive Summary

Twenty-four models of police traffic radar devices were tested for performance characteristics and compliance to the Model Performance Specifications for Police Traffic Radar Devices developed for the National Highway Traffic Safety Administration (NHTSA) by the Law Enforcement Standards Laboratory (LESL) of the National Bureau of Standards (NBS). The specifications were adopted by the International Association of Chiefs of Police (IACP), and the testing program was conducted by them in accordance with a cooperative agreement with NBS/LESL. Testing was accomplished by two independent testing laboratories during the period June 1983 to January 1984.

The radar devices were subjected to examination, including documents and accessories, laboratory evaluation and operational testing. During the period of testing, the equipment manufacturers were afforded the opportunity to make minor modifications to their products to achieve compliance with the model specification requirements. The development of the performance specifications and testing of radar devices to the requirements of the model specification have led to definitive improvements in radar devices that will be available in the future, and to the listing of 24 models on the IACP Consumer Products List (CPL).

It is recommended by the IACP that every agency procuring radar units require the successful bidder to certify that the radar devices are included on the IACP CPL and meet the NHTSA/IACP Model Performance Specifications for Police Traffic Radar Devices. This dual certification is important because some manufacturers have indicated that model numbers on both conforming and nonconforming units may be identical.

Summary of Test Results

(Testing) was accomplished in two phases. Upon completion of the initial testing, representatives of the IACP, NBS and NHTSA reviewed the laboratory test results in detail. Many of the individual radar devices failed to comply with the model specification because the required manufacturer information was not provided with the unit. For example, some devices lacked operating or installation instructions while, for others, the required tuning fork certificate was either incomplete of not provided. In addition, there are a number of instances in the model specifications where the speed measuring radar device is tested to the manufacturer's specification for a given parameter if the manufacturer claims a greater capability than that required by the model specification. For example, the model specification places a limit on signal processing sensitivity variation for targets traveling at speeds of 20 to 90 MPH. In a number of cases, although a radar unit met this requirement of the model specification, the manufacturer data sheet claimed an operating capability below 20 MPH and the unit, when tested at the lower speed, did not meet the signal processing channel sensitivity requirement.

Similarly, there were a number of radar units that did not comply with the display labeling requirements of the model specification. Very few of the radar units provided by the manufacturers were in full compliance with the labeling and the manufacturer-provided information requirements of the model specification at the time of initial testing. However, it was clear from earlier discussion with the manufacturers, in advance of the testing, that the manufacturers were willing to correct differences in the information provided with the units, specifications, and labeling to fully comply with the requirements of the model specification.

There were some areas of noncompliance with the requirements of the model specification that required minor modifications to the individual radar unit to achieve full compliance. Other than labeling and instructions, the most common deficiency was the failure to meet the signal processing channel sensitivity requirements. This deficiency was corrected, in most cases, by a change in the values of the filter components used to control this sensitivity. The second most common deficiency was in the display readability capability of the units, i.e., character height and luminance contrast had to be improved. The design of the radar units is such that either individual components or the display module itself can be easily re-

placed in order to comply with the readability requirements of the model specifications.

Other areas of noncompliance included such items
as frequency stability under conditions of high and low
temperature or high humidity. In examining all of the initial tests' results, it was apparent that the manufacturers
could easily make minor modifications to their products
at a minimal expense to achieve full compliance with the
requirements of the model specification.

The IACP was of the opinion that the law enforcement community would benefit from the wide availability
of speed measuring radar devices that were in full compliance with the requirements of the model specification,
and recommended that all of the manufacturers be given
the opportunity to make those modifications to their products that were necessary to achieve full compliance. The
NHTSA accepted this recommendation, which was also
acceptable to the NBS.

The IACP provided each manufacturer with the
results of the initial test conducted on their products only,
and notified the manufacturers that they would have the
opportunity to correct their product literature and to make
minor modifications to the individual speed measuring
devices to achieve compliance with the model specifications. The conditions of this offer included: 1) any required modification was to be made to the same unit that
had been tested during initial testing, 2) the modification
was to be made at the test facility in the presence of testing
laboratory personnel, 3) the NBS technical staff was to be
provided copies of any circuit modifications so that NBS
could advise IACP as to what additional testing would be
required based on the modifications that were made, and
4) it would be the manufacturer's responsibility to pay for
any required testing following modification.

All five manufacturers accepted the offer. The modifications were made, and the laboratories provided final

phase test results to the IACP. During the course of the testing and subsequent data review by IACP and NBS, all of the deficiencies were corrected to bring the radar devices into full compliance with the model specifications.

Laboratory Tests

Each test is described, and the number of devices in noncompliance is given.

Tuning Fork Calibration (12)

This test requires measurement of the frequency of the tuning fork to ensure that it is within $\pm 0.5\%$ of the frequency specified in the certificate of calibration. This could not be accomplished in the case of 12 radar units as no calibration certificate was supplied by the manufacturer.

Radar Device Tuning Fork Tests (4)

In these tests the tuning fork(s) supplied with a radar unit are used to generate a pseudo-Doppler signal to check the operational condition of the radar unit. If the radar is functioning properly, the speed for which the tuning fork is calibrated should appear in the speed display with a tolerance of ± 1 MPH. With moving radar devices, both target and patrol vehicle displays must show the correct speed readings. The tuning fork tests are conducted under standard test conditions, at specified low and high temperatures, under high humidity conditions and during vibration. All radar devices were in compliance at ambient temperature. However, four devices failed to comply during environmental tests. Three radar units did not function properly at low temperature. One of the three also did not function properly under the high temperature test conditions, and another one of the three failed during the high humidity test. The speed display on one additional unit showed erroneous readings during the vibration testing.

Microwave Transmission

The microwave transmission tests check the stability of operation of the radar device under conditions that could be encountered in the operational environment. These tests were repeated a number of times to record performance data under standard test conditions (68–80 degrees F), under conditions of low temperature (–22 degrees F), high temperature (+ 140 degrees F) and high humidity (90%) with temperature at least 99 degrees F. All tests were repeated at three voltage levels; nominal voltage (13.6V), nominal voltage plus 20 percent (16.3V), and at minus 20 percent (10.8V) or a lower voltage that manufacturer may have specified. The following characteristics were tested as specified above:

(a) Frequency Stability (5)

Five of the radar devices did not maintain frequency stability within the allowable tolerance during these tests. One failed at all conditions, while three others failed the low temperature test. A fifth device failed during the high humidity test.

(b) Input Current Stability (8)

Three of the same radar units that did not meet the frequency stability requirement also did not meet the input current stability requirement of less than a 10 percent variation and no change in numerical display under one or more of the test conditions, two at low temperature and one at high temperature. Three additional units were unstable during the low temperature tests, while a total of four failed at high temperature. The same three plus one other device failed the high humidity tests.

(c) Radiated Output Power Stability (4)

Four radar units did not maintain the radiated power output within the plus or minus 1.5 dB from nominal as required. Two did not comply at low temperature conditions and one failed the high temperature test. That one, plus one other radar device, did not function properly under high humidity conditions.

(d) Antenna Horizontal Beam Width (4)

The maximum horizontal beam width allowed by the model specifications is 18 degrees for X-Band radar and 15 degrees for K-Band. The permissible band width was initially exceeded by three X-Band and one K-Band radar devices.

(e) Antenna Near-Field Power Density (4)

The requirement for this test is that the measured power density may not exceed that specified by the manufacturer. Four radars did not meet this requirement.

Low Voltage Supply (12)

Each radar device is required to operate to a specific low voltage point and to have a low voltage indicator capable of being heard or seen by the operator. The required low voltage operating point is 10.8V or the lowest voltage specified by the manufacturer. At the specified low voltage the speed display must show either no reading or no erroneous reading. Seven radar units did not meet the performance requirements. Five met the requirement of the model specifications, but not the lower voltage specified by the manufacturer.

Doppler Audio

(a) Audio Output and Volume Control (3)

The radar device is required to emit a Doppler audio tone that correlates with the received Doppler signal and to have an audio volume control. The Doppler audio tone is beneficial to the radar operator as an aid in correlating visual observation of a target vehicle with the radar speed display, and in alerting the operator to the presence of interference that may affect his or her ability to operate that radar device. Three radar units were not in compliance.

(b) Audio Squelch and Squelch Override (2)

The audio tone must be squelched in the absence of a target signal. The radar device must also permit the operator to inhibit the squelch action while keeping the receiver open. Two exceptions to this requirement were noted.

(c) Audio Track-Through Lock (1)

For radar devices with a track-through lock feature, the Doppler audio must continue after the speed-lock switch is activated. One radar unit did not meet this specification when initially tested.

Operational Tests

Electromagnetic Interference (3)

The operational tests were conducted with the radar device properly installed in a patrol vehicle of the type normally used for law enforcement purposes. The test vehicle had an FM transceiver and antenna and a citizens band transceiver and antenna, each installed in accordance with the manufacturer's instructions. A handheld FM transceiver was also positioned in the vehicle for use by the driver. While the radar was tracking an acquired target vehicle traveling at 50 MPH, audio tones from 500 to 3000 Hz generated by a slide whistle were transmitted via the microphone of the FM and CB transceivers. The radar display was observed for any erroneous readings caused by the slide whistle transmission from the transceivers. At least two tests were conducted with each transceiver. The tests were repeated similarly with the patrol vehicle, in the stationary mode, tracking a target vehicle while a third vehicle equipped with the FM and CB transceivers passed within 10 feet, first, on one side of the patrol vehicle and then on the other.

One radar device was subject to interference from the FM transceiver when initially tested, while two others were interfered with by the CB transceiver.

Speed Accuracy Tests (5)

The speed accuracy tests were conducted on a half-mile measured course, over which the target vehicle was driven at constant speeds of 20, 50 and 70 MPH. The true speed of the target vehicle for each test was calculated from the elapsed time to travel the known distance, or measured with a fifth wheel speed measuring device. The speed displayed on each radar device during the runs was compared to the true target vehicle speed to determine whether or not the radar-displayed speed was within the allowable variation of + 1, −2 MPH in the stationary mode and ± 2 MPH in the moving mode of operation.

During the initial testing, five radar devices were reported as not complying with the speed accuracy requirements of the model specifications; three in the stationary mode/moving target test and one of these three, in addition to two others, in the moving mode/approaching target test. However, these particular test data are considered questionable, since all five units met the speed accuracy requirements of all other speed tests. The three radar devices that were not in compliance in the stationary mode/moving target test situation at 70 MPH were in compliance when tested at speeds of 20 and 50 MPH in the initial testing phase. Similarly, the three radar devices that did not meet the compliance requirement in the moving mode/approaching target (patrol 20 MPH—target 55 MPH) met the requirement in the higher speed test (patrol 55 MPH—target 70 MPH), and also the stationary tests at speeds of 20, 50 and 70 MPH. All five units were found to comply fully with the speed accuracy requirements when retested during the final phase of testing. It should be noted that the final testing was conducted without adjustment or modification of the speed measuring circuitry of any of the five devices.

TESTS OF SIX SPEED MEASURING RADAR UNITS: NATIONAL BUREAU OF STANDARDS

Submitted to National Highway Traffic Safety Administration by Law Enforcement Standards Laboratory, January 1980, National Bureau of Standards.

Executive Summary

E ach of the six radar models identified in hearings conducted by the Eleventh Judicial Circuit Court of Florida were subjected to a series of laboratory and field tests.

The tests were designed to determine whether it was possible to affect target speed measurement under those environments or use situations discussed in the Florida hearing.

There was no observed degradation in the performance of any of the units arising from 1) cosine angle effect, 2) automatic lock, 3) heat build-up, 4) mirror switch aiming, and 5) high tension power wire interference. Only one of the six units tested was affected by police radio transmission at distances beyond 30 feet from the radar.

Certain of the units were affected in varying degrees by internal electrical interferences from the ignition and alternator, and interference from air conditioner and heater fans. While transmission from police radios in the same vehicle as the radar was found to have a limited effect on the radar units, CB transmission in the same vehicle affected nearly all radars.

Patrol car shadowing was found to affect all but one of the radars, and target speed bumping was observed to affect half of the units. In all cases, the two-piece radars were affected by panning the antenna beam through the display console.

In some instances, the performance of the radar units was not degraded by the environments or use situations described above when a target vehicle was present.

The limited sample size precludes generalization of the test results to an entire manufacturer product line. While there is no question that the individual radar units that were tested are affected to varying degrees by interference sources and certain operating conditions, it appears, based upon observation, that the source of error can be avoided by taking precautions in the installation of the radar unit and through proper use by a skilled and knowledgeable operator.

Test Results

The test results for each operational situation or environment as identified in the Florida hearing are presented separately in the paragraphs that follow.

Shadowing

Shadowing, better identified as patrol speed shadowing, is the tendency of a moving mode radar to use a slow moving large vehicle rather than the ground to measure the speed of the patrol car. This effect was observed, in varying degrees, for several of the radars during testing.

Radar Unit A (Kustom MR-9): Patrol speed reading went from an actual 26 MPH to indicated 22 MPH as large oil tanker passed patrol car. No other effects from large trucks were observed.

Radar Unit B (MPH K-55): Large truck came to a stop 50 yards in front of patrol car and the patrol car reading changed from 40 to 28 MPH. Several large trucks traveling in front of the patrol car and trucks passing patrol car had no effect on readings.

Radar Unit C (Decatur MV-715): Actual patrol speed of 25 reduced to 17 when following a large truck. Actual patrol speed of 34 reduced to 11 MPH for a long time when a pickup truck passed the patrol car. Pickup passing patrol car decreased patrol speed reading by 10 MPH and increased target speed reading by 10 MPH.

Radar Unit D (CMI Speedgun 6): This particular unit demonstrated very severe shadowing effects from moving vehicles in the vicinity of the radar. A 44 MPH patrol speed momentarily dropped to 11 MPH with a pickup truck 100 feet in front of the patrol car. While traveling at actual patrol speeds of 30 to 40 MPH and target speeds of 50 to 60 MPH, the following readings were observed as the patrol car was passed by trucks.

Patrol	Target
12	80
16	72

20	57
17	68

This unit shadowed most trucks, large or small.

Radar Unit E (CMI Speedgun 8): While traveling at patrol speed of 50 MPH, shadowing from passing large trucks caused radar reading of:

Patrol	Target
17	74
11	41

This effect did not occur every time a truck passed. Patrol readings decreased momentarily several times when a passenger sedan passed patrol car.

Radar Unit F (MPH K-55): While the patrol car was traveling 35 MPH, readings of 20 patrol and 76 target were observed. Then a pickup passed the patrol car and pulled back into the lane. Actual patrol speed of 47 MPH dropped to 35 MPH as a large truck passed patrol car. This happened about 20 percent of the time that trucks passed.

Radar Unit G (Kustom MR-7): No effects on target readings or patrol speed readings of shadowing from passing large trucks were detected.

Batching

Batching, better identified as target speed bumping, occurs if the target vehicle speed display varies when the patrol car changes speed. This effect was observed for some of the radars during testing.

Radar Unit A (Kustom MR-9): No effect on target speed readout due to acceleration or deceleration of patrol car.

Radar Unit B (MPH K-55): Radar occasionally subtracted 2 to 3 MPH from target reading when patrol car was accelerated. This was not consistent. No effect on deceleration.

Radar Unit C (Decatur MV-715): Accelerating or decelerating patrol car had no effect on target readings.

Radar Unit D (CMI Speedgun 6): Radar target speeds decreased with both acceleration and deceleration. Target speed decreased 3 MPH when patrol speed suddenly increased. Target speed decreased 5 MPH when patrol speed suddenly decreased.

Radar Unit E (CMI Speedgun 8): No effect on target speed reading could be observed from accelerating or decelerating patrol car.

Radar Unit F (MPH K-55): On rare occasions target speed readings increased 2 or 3 MPH when speed of patrol car was increased or decreased. Mostly no effect.

Radar Unit G (Kustom MR-7): No effect observed on target readings from sudden increase or decrease of patrol car speed.

Panning Error

Panning error occurs when the radar antenna is used to pan through its own display. The two one-piece units cannot do this. Using the radar in this fashion always produced an erroneous reading.

Radar Unit A (Kustom MR-9): Panning antenna across radar readout console produced 101 MPH readings.

Radar Unit B (MPH K-55): Panning antenna across radar readout console produced 75 MPH readings.

Radar Unit C (Decatur MV-715): Panning antenna across radar readout console caused readings over 100 MPH.

Radar Unit D (CMI Speedgun 6): NA.

Radar Unit E (CMI Speedgun 8): NA.

Radar Unit F (MPH K-55): Panning antenna across radar readout console produced various readings.

Radar Unit G (Kustom MR-7): Panning antenna across radar readout console produced 95 MPH speed readings.

Scanning Error

Scanning errors can result when the radar operator moves his antenna too quickly. The results obtained from testing by positioning the radar in commonly used positions or pointing in several directions are discussed below. Although erroneous readings were observed for some of the radars during these tests, this interference was not observed when the radars were tracking a target during operational testing.

Radar Unit A (Kustom MR-9): No effect from AC/heater fan motor or AC/heater fans when pointed any direction inside the car.

Radar Unit B (MPH K-55): Picked up AC/heater fan intermittently while mounted on dash. When heater fans were on, scanning dash with handheld antenna caused readings proportional to fan speed. No readings from fans when antenna was pointed out the side windows. Picked up fan readings when radar was pointed out rear window (radar held above back of front seat cushion).

Radar Unit C (Decatur MV-715): When mounted, antenna pointed slightly downward toward dash with heater fans on high: Stationary Mode—target 34, patrol 55. No readings from fans when antenna pointed out the side or rear windows. Antenna mounted outside rear passenger window slightly aimed at dash with fans on, picked up readings 24 to 25 MPH in stationary mode.

Radar Unit D (CMI Speedgun 6): Handheld radar pointed at dash, picked up AC/heater fans. No effect from fans, no readings when pointed out the side or rear windows.

Radar Unit E (CMI Speedgun 8): Handheld radar antenna pointed at dash picked up readings from AC/heater fans. This unit seemed to pick up the fans extremely easily when antenna was aimed forward but not mounted on dash mount. No readings when antenna was aimed out the side or rear windows.

Radar Unit F (MPH K-55): No fan readings when unit was dash mounted. When radar antenna is pointed at dash, radar picks up fan readings. No speed readings when antenna is pointed out the side or rear windows.

Radar Unit G (Kustom MR-7): Scanning dash with radar antenna did not pick up AC/heater fans. No readings when antenna was pointed out the side or rear windows with all fans on.

Cosine Angle Effect

The cosine error, better named cosine angle effect, was closely observed during the testing, for this effect is present any time that the radar and the target vehicle are not heading directly toward each other. The speed the radar displays when in the stationary mode, is equal to the actual speed times the cosine of the angle between the direction of travel of the target vehicle and the radius vector from the radar to the target vehicle. No erroneous readings were noted during these tests due to the cosine angle effect.

Automatic Lock

The advantages and disadvantages of the automatic lock feature were also observed. The advantage is in enabling the officer to automatically obtain a speed on the display console using a present threshold value. The disadvantages far outweigh this advantage. The primary one is that it prevents the radar operator from obtaining a tracking history, i.e., what the vehicle being tracked does after it is detected traveling over the speed limit. Also, the radar may automatically lock on to a stray abnormal reading which only appeared momentarily due to an unobserved interference source. Another disadvantage is that use of the automatic lock feature makes it almost impossible to compare the patrol car speed with its radar measured counterpart to insure that the two agree at the time the display is locked in or at times of possible interference such as that brought on by patrol speed shadowing. Again, there was no separate test to observe the use of the automatic lock feature.

External Interference

Theoretically, defective high tension wires may affect radar speed measuring devices by causing stray readings or by decreasing the sensitivity of the radar and thereby reducing its range of operation. However, no stray readings were noted in testing near and under high tension wires for any of the models tested.

Four of the radars did not display false readings due to external police radio transmission and only one radar was affected by external CB operation.

Radar Unit F (MPH K-55): CB radio mounted in a pickup caused readings of 60 to 70 MPH at distances up to 175 feet from patrol car. The 100 watt transmitter. . .caused radar readings at 5 feet from the left rear of the patrol vehicle to 30 feet to the right of the patrol vehicle.

Internal Interference

Radars were operated in their usual configuration in a patrol car and the electrical interference due to the patrol car ignition system, alternator, and air conditioner and heater fan motors was investigated. Some of the radar units displayed readings when the air conditioner and heater fans were operated with no radar target present, and one type displayed a reading during the ignition and alternator testing.

Radar Unit C (Decatur MV-715): With radar mounted in pickup, erratic patrol speed readings from engine alternator and/or ignition in the moving mode. Extreme interference from heater fan motor (electrical) in the moving mode.

When a CB radio was installed in the same vehicle as the radar unit, operating from the same battery, several of the radar units were subject to interference during CB transmission.

Radar Unit A (Kustom MR-9): When the CB radio and the radar were mounted in the same truck and connected to the same battery, interference from the CB radio was evident. Interference from the tone whistled into the microphone caused speed readings to be displayed.

Radar Unit B (MPH K-55): When radar and CB radio were connected to the same battery in truck, readings were observed on radar target readout depending on frequency of CB radio.

Radar Unit C (Decatur MV-715): Radar and CB radio connected to same battery in pickup caused readings in the stationary mode.

Radar Unit D (CMI Speedgun 6): Interference from CB radio when radar and CB radio were powered by same battery supply.

Radar Unit E (CMI Speedgun 8): No effect when radio and radar were both connected to the same battery in the pickup.

Radar Unit F (MPH K-55): Not tested because the unit was found extremely sensitive to external CB transmission.

Radar Unit G (Kustom MR-7): Radar and CB radio connected to same battery in pickup caused reading on radar target display. Under these conditions, radar picked up correct reading on passing target car and whistling in CB microphone had no effect on correct reading until target was out of range. Two of the radars were affected by operating police radio in the same car.

Radar Unit D (CMI Speedgun 6): 100 watt transmitter had no effect on large targets or strong signals, but it boosted target speeds as much as 20 MPH on distant weak signals or small targets.

Radar Unit G (Kustom MR-7): 100 watt transmitter caused increases or decreases of 10 MPH in target speed readings intermittently.

Observations

1. Two-piece radars can produce erroneous readings when an antenna is panned through the display console. The radar should not be mounted with the display console in the antenna beam.

2. Air conditioner and heater fans and alternator or ignition noise can interfere with the radar when no bona fide radar target is present. The radar antenna should be mounted so that it is not pointing toward air conditioner or heater fans. If possible, the antenna should be mounted outside of the patrol vehicle.

3. Patrol speed shadowing can occur during moving mode radar operations. Operators should be aware of this and recognize its symptoms, should know its cause and that its effect can best be detected by checking the radar patrol car speed with the patrol car speedometer.

4. Target speed bumping can occur during moving mode radar operations. Operators should know what it is and how it occurs and that its effect can be avoided by maintaining constant speed when making speed measurements.

5. The cosine angle effect can occur during stationary radar operation when the target is off axis within the antenna beam. Operators should understand the cosine angle effect and recognize when it is occurring.

6. Transmission from 100 watt FM police radios, both external to and within the patrol vehicle, can cause the radar display to blank or produce erroneous readings. Operators should be aware of this and not transmit on police radios while using the radar.

7. Internal transmission from CB radios can produce erroneous readings. Operators should be aware of this and not transmit on CB radios while using the radar.

8. The use of the automatic-lock feature may result in wrong target identification. Operators should be aware of this and not use the automatic-lock feature, but instead use the manual lock feature and then only when they have observed sufficient "tracking history" to insure that the correct target vehicle is being tracked.

9. When two or more vehicles are in the radar beam, it can be difficult to select the correct target. Operators should continue radar tracking until the proper target is positively identified; it may be necessary to wait until the vehicles pass by the patrol car and the one being tracked no longer registers a speed on the display console.

ENDNOTES

1. *A Comparison of the Automobile Accident Rates of Radar Detection Device Users and Nonusers,* Yankelovich, et. al., 1987.
2. *Accidents on Main Rural Highways Related to Speed, Driver and Vehicle,* David Solomon, FHwA, 1964.
3. *Traffic Control and Roadway Elements,* Donald E. Cleveland, University of Michigan, 1970.
4. *The Effects of the 65 MPH Speed Limit During 1987,* National Highway Traffic Safety Administration, January 1989.
5. *Interstate System Accident Research Study, Interim Report II,* "Public Roads," Julia Anna Cirillo, FHwA, August 1968.
6. *55: A Decade of Experience,* Report of the Transportation Committee, National Academy of Sciences, 1984.
7. *Ibid.*
8. *Interstate System Accident Research Study.*
9. *Selective Traffic Enforcement Program,* report to the National Highway Traffic Safety Administration, NTIS PB-257-430, August 1976.
10. *Report to the House Transportation Subcommittee,* Kenneth M. Mead, General Accounting Office, February 1988.
11. *Radar as a Speed Deterrent: An Evaluation,* Donald W. Reinfurt, et. al., Highway Safety Research Center, University of North Carolina, February 1973.
12. *The Ticket Book,* Rod Dornsife, Bantam Books, 1980.
13. *Col. Lee Nichols,* Dean of Department of Engineering, Virginia Military Institute, in "Radar on Trial," videocassette, 1986.
14. *Spurious Signals,* John Tomerlin, "Road & Track," May 1981.
15. *Ibid.*
16. *What Should the Maximum Speed Limit Be?,* Matthew C. Sielski, "Traffic Engineering," September 1956.
17. *Ibid 2, 4.*
18. *Is Traffic Radar Reliable?,* National Highway Traffic Safety Administration, DOT-HS-805-254, February 1980.
19. *Testing of Police Traffic Radar Devices,* International Association of Chiefs of Police, April 1984.

INDEX

Disclaimer

This book gives the reader a guide to basic procedures in traffic courts and the prosecution and defense of speeding citations. The procedures may vary from state to state or court to court. This book does not qualify the reader to practice law and the reader is not to place any reliance on this book in that regard.

The authors are not, by furnishing the information contained in this book, acting as the reader's attorneys or representing him/her/them in any way. Questions concerning proper practice of law or procedure and/or interpretation of any statute, ordinance, regulation or case law should be referred to a qualified attorney in the reader's jurisdiction.